应用型本科机械工程系列精品教材

English for Die & Mould Design and Manufacturing

模具设计与制造专业英语

王家惠　郑兴睿　主　编

万祥明　吴承玲　程　良　副主编

清华大学出版社

北京

内 容 简 介

本书是为适应高等学校模具设计与制造专业和机械工程及自动化专业模具方向专业英语教学需要而编写的。本书从模具专业特点出发,并注意吸收现代模具新技术方面的有关知识,涵盖了模具设计与制造的主要内容,包括:冷冲模、塑料模、压铸模、数控加工、模具特种加工、模具 CAD/CAE/CAM 等。为便于阅读,在编排上各单元相对独立,每节后附有新出现的专业词汇,同时在书后附有模具专业词汇总表。

本书可作为高等学校及高职教育模具设计与制造专业英语教材,也可供有关工程技术人员参考,或作为工具书使用。

图书在版编目(CIP)数据

模具设计与制造专业英语/王家惠,郑兴睿主编. —北京:清华大学出版社,2017
(应用型本科机械工程系列精品教材)
ISBN 978-7-302-48065-5

Ⅰ. ①模… Ⅱ. ①王… ②郑… Ⅲ. ①模具－设计－英语－高等学校－教材 ②模具－制造－英语－高等学校－教材 Ⅳ. ①TG76

中国版本图书馆 CIP 数据核字(2017)第 208130 号

责任编辑: 赵　斌　赵从棉
封面设计: 常雪影
责任校对: 王淑云
责任印制: 沈　露

出版发行: 清华大学出版社
网　　址: http://www.tup.com.cn, http://www.wqbook.com
地　　址: 北京清华大学学研大厦 A 座　　**邮　　编:** 100084
社 总 机: 010-62770175　　**邮　　购:** 010-62786544
投稿与读者服务: 010-62776969, c-service@tup.tsinghua.edu.cn
质量反馈: 010-62772015, zhiliang@tup.tsinghua.edu.cn
印 装 者: 三河市少明印务有限公司
经　　销: 全国新华书店
开　　本: 185mm×260mm　　**印　　张:** 9.25　　**字　　数:** 221 千字
版　　次: 2017 年 12 月第 1 版　　**印　　次:** 2017 年 12 月第 1 次印刷
印　　数: 1～1000
定　　价: 42.00 元

产品编号:070494-01

前言

FOREWORD

21 世纪我国将成为制造业强国，与国外交往的机会亦会随之增多，对外学术交流及合作将更加频繁，以英语为载体的专业信息将成倍增长，专业英语的应用亦将越来越广泛。对模具设计与制造专业的研究生、本科学生及专科学生，以及从事模具相关行业的工程技术人员而言，熟练掌握专业英语，对于促进国际交流、了解并熟悉国外模具设计与制造行业的最新发展动态、参与模具产品的国际竞争是十分必要的。模具设计与制造专业英语是模具设计与制造专业的一门重要基础课。随着模具 CAD/CAM/CAE 技术的发展和互联网的普及，新世纪从事模具设计与制造专业的毕业生不仅要掌握先进模具技术，而且要具备扎实的外语能力。模具设计与制造技术的发展将对模具专业英语教学提出更高的要求，对专业英语的学习亦将更为迫切。为了满足模具设计与制造专业英语教与学的需求，我们编写了《模具设计与制造专业英语》一书。

模具设计与制造是一门交叉学科，内涵丰富，涉及面很广，包括金属材料成形、高分子材料成形、模具材料、模具制造工艺、先进制造方法等内容。在整个编写过程中，为使本教材体现先进性、科学性和实用性，本书从国外最新出版的教科书、专著、外文期刊中筛选资料，把国外较新的研究成果编写进教材中；并根据该课程教学大纲要求，从培养学生阅读能力方面着手，充分考虑了文章的阅读性与知识性，所选资料既考虑了当今模具行业的覆盖面，又反映了其发展趋势。在侧重阅读理解、掌握模具专业常用词汇基础上，突出模具专业特点。课文简单易读，适合不同层次的读者阅读。

本书的选材是在有限的篇幅内尽可能地涵盖模具设计与制造的学科领域。全书共分 6 章，即冲压工艺及模具设计（Stamping Forming and Die Design）、塑料模具设计（Plastics Molds）、铸造模具设计（Casting Dies）、数控加工（CNC Machining）、模具特种加工（Mold Special Machining）以及模具 CAD/CAM/CAE。

本书由昆明理工大学王家惠、郑兴睿任主编。其中王家惠负责全书的结构及第 2 章，第 5 章内容的编写；郑兴睿负责第 1 章，第 4 章内容的编写；吴承玲负责第 3 章内容的编写；程良负责第 6 章内容的编写；万祥明参加了第 5 章内容的编写；书中插图由万祥明绘制。

本教材在编写过程中，得到了昆明理工大学模具技术研究所的教师和研究生的大力支持，在此表示衷心的感谢。

由于时间仓促，编者水平有限，书中错误之处在所难免，敬请批评指正。

编　者

2017 年 7 月

目录

CONTENTS

Chapter 1

Stamping Forming and Die Design

1.1 Introduction

In today's practical and cost-conscious world, sheet-metal parts have already replaced many expensive cast, and machined products.

The reason is obviously the relative cheapness of stamped, or otherwise mass-produced parts, as well as greater control of their technical and aesthetic parameters. That the world slowly turned away from heavy, ornate, and complicated shapes, and replaced them with functional, simple, and logical forms only enhanced this tendency. Remember old bathtubs ? They used to be cast and had ornamental legs. Today they are mostly made of coated sheet metal, if not plastics. Manufacturing methods for picture frames, and doors were gradually replaced by more practical and less costly techniques.

Metal stamping, probably the most versatile products of modern technology, are used to replace parts previously welded together from several components. A well-designed sheet-metal stamping can sometimes eliminate the need for riveting or other fastening processes. Stampings can be used to improve existing designs that are often costly and labor-intensive. Even products already improved upon, with their production expenses cut to the bone, can often be further improved, further innovated and further decreased in cost.

This part focuses on the stamping forming technology and its die design in metal processing. Stamping is mainly used in sheet plate forming, which can be used not only in metal forming, but also in non-metal forming. In stamping forming, under the action of dies, the inner force deforming the plate occurs in the plate. When the inner force reaches a certain degree, the corresponding plastic deformation occurs in the blank or in some region of the blank. Therefore the part with certain shape, size and characteristic is produced.

Stamping is carried out by dies and press, and has a high productivity. Mechanization and automatization for stamping can be realized conveniently owing to its easy operation. Because the stamping part is produced by dies, it can be used to produce the complex part that may be manufactured with difficulty by other processes. The stamping part can be used generally without further machining. Usually, stamping process can be done without heating. Therefore, not only does it save material but also energy. Moreover, the stamping part has the characteristics of light weight and high rigidity.

Stamping processes vary with the shape, the size and the accuracy demands, the output of the part and the material. It can be classified into two categories: cutting process and forming process. The objective of cutting process is to separate the part from blank along a given contour line in stamping. The surface quality of the cross-section of the separated part must meet a certain demand. In forming processes, such as bending, deep drawing, local forming, bulging, flanging, necking, sizing and spinning, plastic deformation occurs in the blank without fracture and wrinkle, and the part with the required shape and dimensional accuracy is produced.

The stamping processes widely used are listed in Table 1-1.

Table 1-1 Classification of the Stamping Processes and Their Characteristics

Process		Diagram	Characteristics
Cutting	Shearing	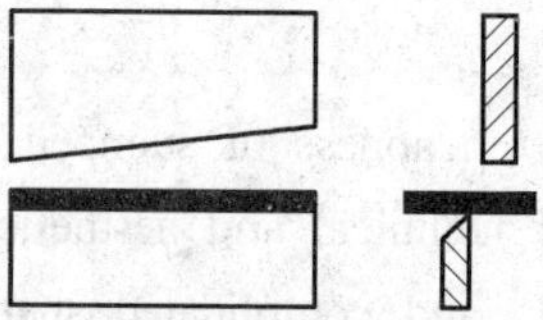	Shear the plate into strip or piece
	Blanking	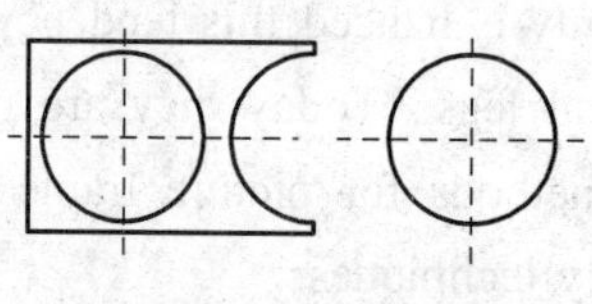	Separate the blank along a closed outline
	Lancing	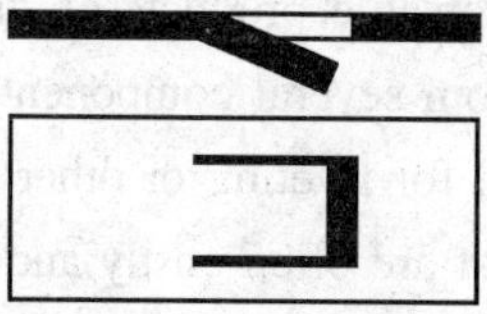	Partly separate the blank along a unclosed outline, bending occurs at the separated part
	Parting	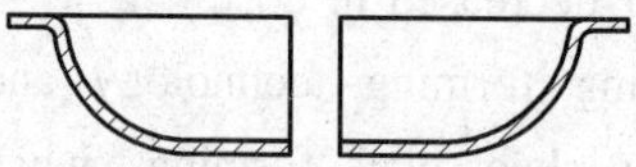	Separate various workpiece produced by stamping into two or more parts
	Shaving	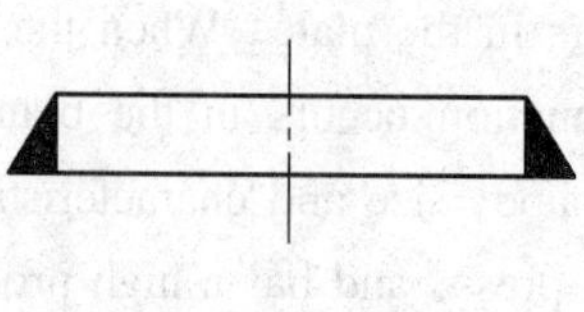	A layer of thin chip is shaved along the external side or the inner hole, to improve size accuracy and smoothness of the cross section of shearing

Continued

Process		Diagram	Characteristics
Forming	Bending	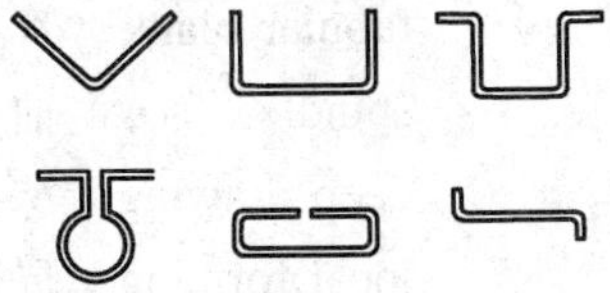	Press the sheet metal into various angles, curvatures and shapes
	Curling	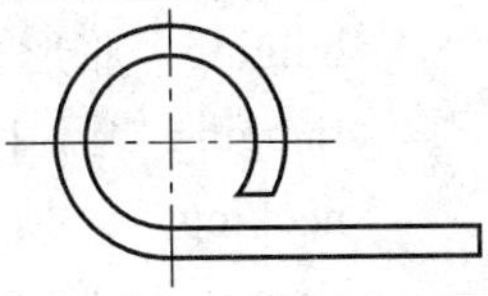	Bend ending portion of the plate into nearly closed circle
	Deep drawing	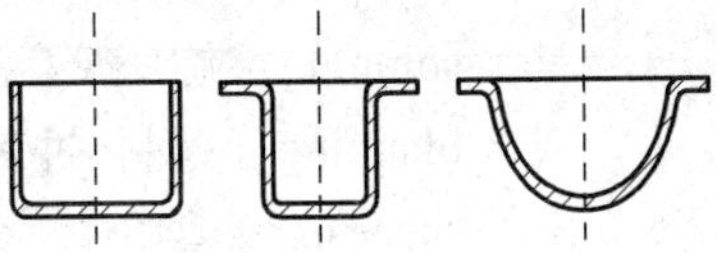	Produce an opened hollow part with punch and die
	Local forming	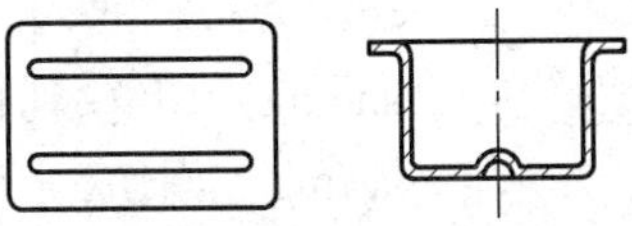	Manufacture various convex or concave on the surface of the plate or part
	Bulging	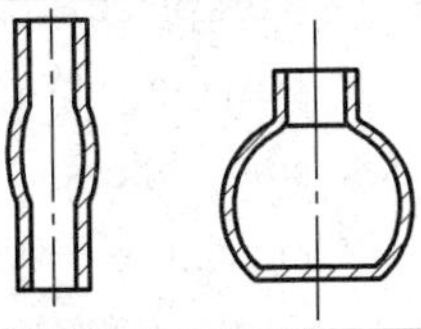	Expand a hollow or tubular blank into a curved surface part
	Flanging	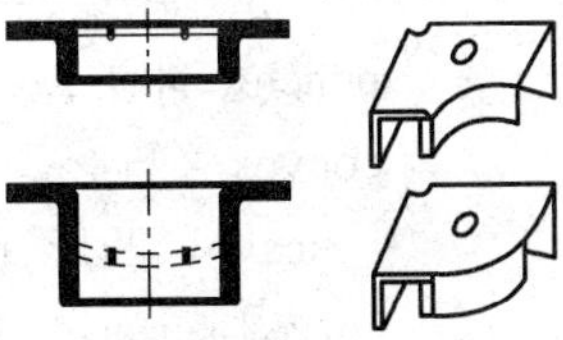	Press the edge of the hole or the external edge of the workpiece into vertical straight wall
	Necking	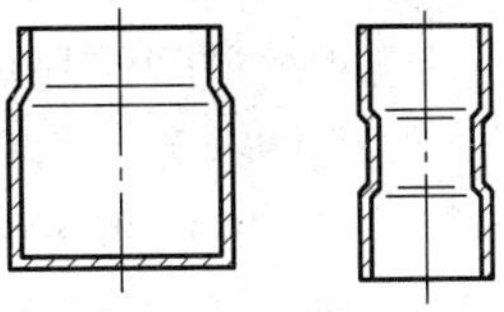	Decrease the end or middle diameter of the hollow or tubular shaped part
	Sizing	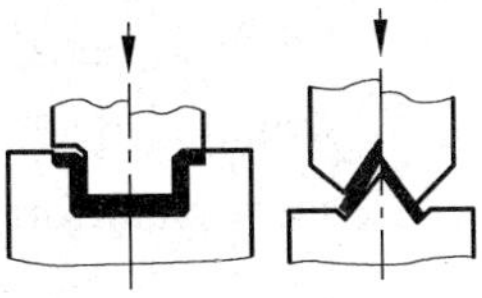	Finish the deformed workpiece into the accurate shape and size
	Spinning	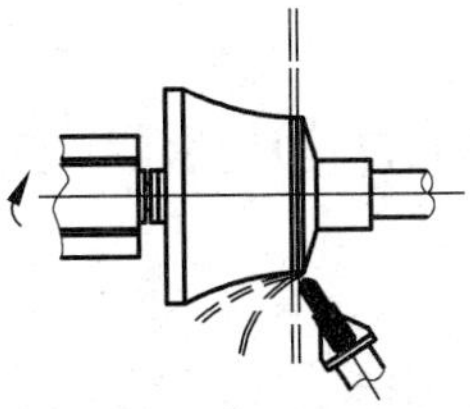	Form an axis-symmetrical hollow part by means of roller feeding and spindle rotational movement

New Words and Expressions

stamping 冲压,冲压件
sheet-metal parts 钣金零件
aesthetic 工艺的
ornate 华丽的
functional 功能的
bathtub 浴盆
ornamental 观赏的
coated sheet metal 喷涂的板材
welding 焊接
riveting 铆接
fastening 紧固
labor-intensive 劳动密集的
die 模具,砧子,凹模
metal processing 金属加工
metallic engineering science 金属工程科学
plastic forming 塑性成形
machining 机械加工,切削加工
casting 铸造,浇注
plate 板,板材,钢板
blank 毛坯,坯料
press 冲压,压制
mechanization 机械化
automatization 自动化
rigidity 刚性,刚度
roller feed 滚轮送料
cutting process 分离工序
forming process 成形过程,成形工艺
contour 轮廓,外形
cross-section 横截面
bending 弯曲
tubular-shaped 管状的
tubular blank 管状坯料
spindle 轴,主轴
deep drawing 深拉延
local forming 局部成形
bulging 胀形,起凸
flanging 翻口,翻边,弯边
necking 缩颈
sizing 整形,矫正
spinning 旋压
blanking 落料,冲裁
shearing 剪切
strip 条料,带料,脱模
lancing 切缝,切口
parting 剖切,分开,分离
workpiece 工件
shaving 修边,整修
smoothness 光滑(度),平整(度)
curvature 弯曲,曲率
curling 卷边,卷曲
punch 冲头,冲孔
convex 凸形
concave 凹形,凹面
curved-surface 曲面
axis-symmetrical 轴对称的
accuracy 精度
engineering science 工程科学
plastic deformation 塑性变形
fracture 断裂,断裂面
wrinkle 起皱
sheet plate forming 板料成形

1.2 Blanking and Punching Dies

1.2.1 Blanking

Metal cutting is a process used for separating a piece of material of predetermined shape and size from the remaining portion of a strip or sheet of metal. It is one of the most extensively used processes throughout die and sheet-metal work. It consists of several different material-parting operation, such as blanking, punching, trimming, parting and shaving, where punching and blanking are the most widely used.

Blanking and punching are the processes to separate sheet metal along a closing outline. After blanking and punching, the plate is separated into two parts. Punching is to punch a needed hole in a blank or workpiece, and the material punched from the blank is the waste, that is, the part out of the closing outline is the workpiece, and the part in the closing outline is the waste. Oppositely, blanking is to punch a workpiece or blank with needed shape in the plate, that is, the part in the closing outline is the workpiece. The part out of the closing outline is the waste. The deformation process and the die structure are identical in both blanking and punching. Conventionally, both blanking and punching are called blanking. Through blanking process, final product as well as semi-finished product for other forming process can be produced.

In the case of the cushion ring shown in Fig. 1-1, the process to make the circle of ϕ22 mm is called blanking, and that to make the inside hole of ϕ10.5 mm is called punching.

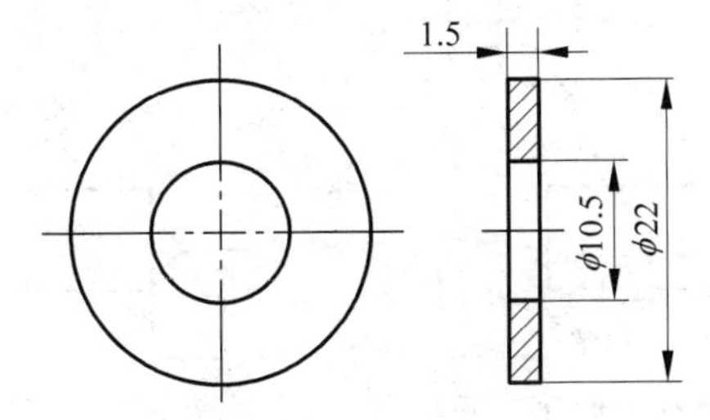

Fig. 1-1 Cushion ring

The actual task of cutting is subject to many concerns. The quality of surface of the cut, condition of the remaining part, straightness of the edge, amount of burr, dimentional stability, all these are quite complex areas of interest, well known to those involved in sheet-metal work.

With correct clearance between the punch and die, almost perfect edge surface may be obtained. This, however, will drastically change when the clearance amount increases, and a production run of rough-edged parts with excessive burrs will emerge from the die.

Highly ductile materials, or those with greater strength and lower ductility, lesser thicknesses or greater thickness—these all were found similarly susceptible to the detrimental effect of greater than necessary clearances.

Naturally, a different type of separation must occur with a softer material than with its harder counterpart. The carbon content certainly has an influence on this process as well. Therefore, the tolerance range must have a provision to change not only with the stock thickness but with its composition as well.

As already mentioned, good condition of tooling is absolutely essential to the cutting process. We may have the most proper tolerance range between the punch and die, and yet the cut will suffer from imperfections if worn-out tools are used.

1.2.2 Blanking Deformation Process

A blanking process involves placing the blank on the die, moving the punch downward to deform and separate the blank with the edges of the punch and die. A clearance Z is existed between the punch and die. The forces of the punch and die applying on the blank are mainly concentrating on the edges of the punch and die.

Blanking deformation process is shown in Fig. 1-2. Under the actions of the punch and die with sharp cutting edges and an appropriate clearance, deformation process undergoes three stages, namely, elastic deformation, plastic deformation and fracture separating stages.

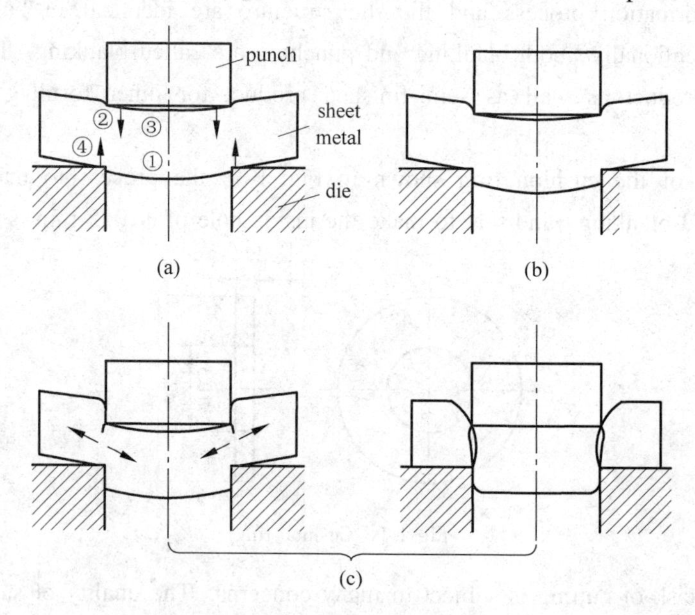

Fig. 1-2 Deformation process of stamping

(a) elastic deformation stage; (b) plastic deformation stage; (c) fracture separating stage

1. Elastic Deformation Stage

When the punch contacts the blank, the material is compressed, resulting in tensile and bending elastic deformation. In this stage, the inner stress hasn't exceeded the elastic limit of the blank yet. The deformation would recover if unloading is occurred.

2. Plastic Deformation Stage

When the punch presses further downward on the blank, the inner stress of the blank reaches its yield strength, the plastic flow and sliding deformation begin to occur. Under the pressure of the punch and die, the surface of the blank is subjected to compression, due to the clearance between the punch and die, the blank is subjected to the actions of bending and tension simultaneously, the material beneath the punch is bended, and that above the die is curled upwards. Circular angles are formed in regions ① and ② due to bending and tension, and indentations appear in regions ③ and ④. While the punch squeeze further into the blank, the plastic deformation and the work hardening in the deformation zone increase further. When the inner stress of the blank near the cutting edge reaches the strength limit of the material, the blanking force reaches its maximum and the cracks occur in the blank, resulting in the damage of the material and the end of the plastic deformation stage (see Fig. 1-2).

3. Fracture Separating Stage

With the punch squeezing into the blank continuously, the cracks at the top and bottom extend to the inner layer of the sheet metal gradually, when the two cracks meet, the blank is cut, and then the process of fracture is ended.

Equilibrium of forces in the shearing zone during blanking is shown in Fig. 1-3; where F_1 and F_2 are the acting forces of the punch and die perpendicular to the blank respectively; F_3 and F_4 are the lateral pressures of the punch and die exerting on the blank respectively; μF_1, μF_2 are the frictions on the end surfaces of the punch and die acting on the blank respectively; μF_3, μF_4 are the frictions on the lateral surfaces of the punch and die acting on the blank respectively. The directions of μF_1 and μF_2 vary with the clearance between the punch and die.

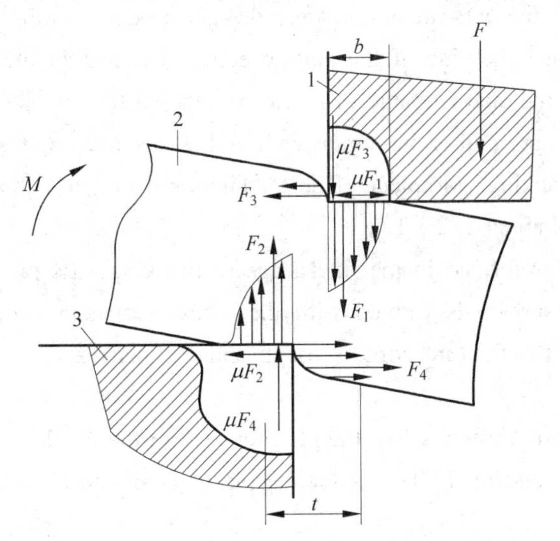

Fig. 1-3 Diagram of the blanking force

1-punch; 2-blank; 3-die

Analysis of the blanking forces shows that the lateral pressures F_3 and F_4 must be smaller than the perpendicular pressures F_1 and F_2; and that the cracks occur and extend more easily in the area of small pressure. Therefore, the initial crack occurs on the side surface of the die in blanking. Observation on crack initiating and developing with scanning electronic microscope shows that when the depth of punch squeezing downward into the material reaches 20% of the blank thickness, the crack occurs on the side surface of the punch and die edges, and then cracks at the top and bottom extend rapidly. When the two cracks meet, the blank is sheared and the process of fracture is ended.

1.2.3 Blanking Workpiece Quality

The quality of the blanking workpiece mainly refers to the qualities of the cutting cross-section and workpiece surface, shape tolerance and dimensional accuracy. The cutting cross-section quality of the workpiece is an important factor to determine whether the blanking process is succeeded or not.

As shown in Fig. 1-4, the cutting cross-section can be divided into four regions: the smooth sheared zone, fracture zone, rollover zone and burr zone.

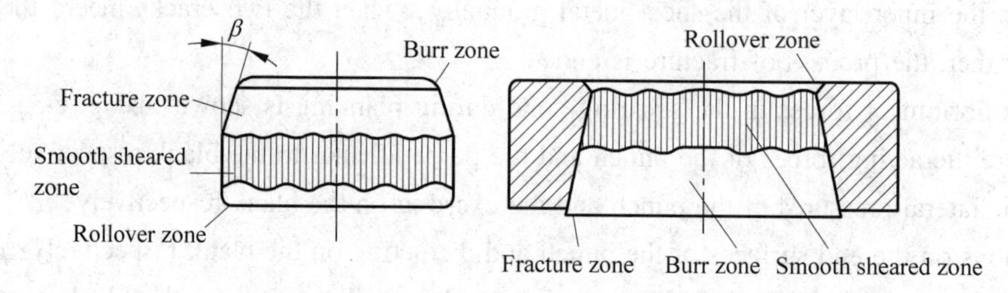

Fig. 1-4 Characteristic of the cutting cross-section of blanking workpieces

When the punch edge cuts into the blank, the plastic deformation occurs due to extrusion between the material and the side of the cutting edge, resulting in the forming of the smooth sheared zone. Due to the characteristic of extrusion, the surface of the smooth sheared zone is smooth and perpendicular, and is the region with highest accuracy and quality within the cutting cross-section of the blanking workpiece. The thickness ratio of the smooth sheared zone to the cutting cross-section is about 1/2 ~ 1/3.

The fracture zone is formed in the final stage of blanking, it's the area where blank is cut off, and the fracture surface is formed with the cracks expanding continuously under tensile stress. The surface of the fracture zone is rough and inclined, and is not perpendicular to the blank.

The rollover zone is formed when the die presses into the blank. The material near cutting edge is embroiled and deformed. The better the plasticity of the material, the larger would be the rollover zone.

The burr of the cutting cross-section is formed when micro-cracks occur during blanking. The formed burr is then elongated and remains on the workpiece.

There are many factors affecting the quality of the cutting cross-section. The proportion of the thickness of the four zones (smooth sheared zone, fracture zone, rollover zone and burr zone) varies with blanking conditions, such as workpiece material, punch and die, equipment, etc.

Fig. 1-5 shows the main factors that affect the quality of the cutting cross-section of blanking workpiece. Fig. 1-6 shows those factors affecting the dimensional accuracy of blanking workpiece. The research and analysis show that the clearance between the punch and die is the most important factor affecting the surface quality and the dimensional accuracy of the blanking workpiece. To increase the surface quality of the blanking workpiece, it is important to study the clearance influence mechanism, so as to find a method for calculating the optimal clearance between the punch and die.

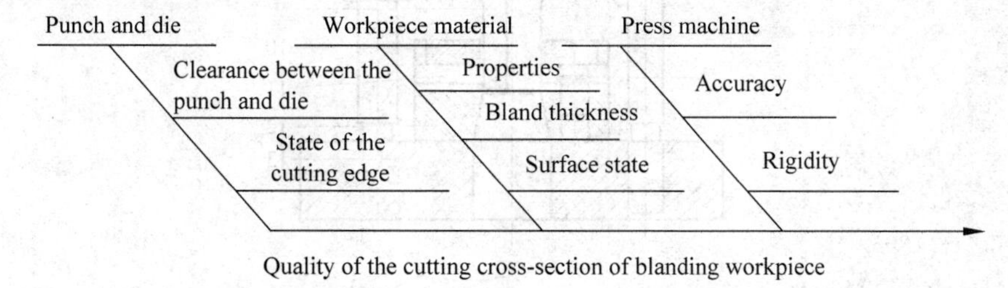

Fig. 1-5 Factors affecting the quality of the cutting cross-section of blanking workpiece

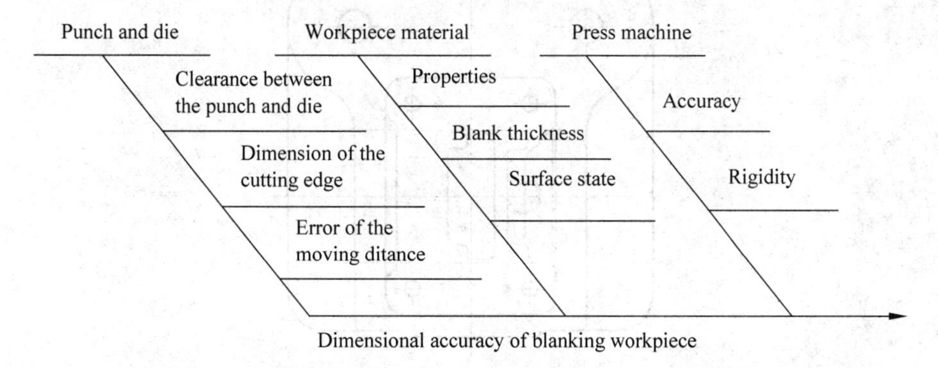

Fig. 1-6 Factors affecting the dimensional accuracy of blanking workpiece

1.2.4 Blanking and Punching Dies

1. Typical Structure of Blanking Die

(1) Simple Die

The die that only one process is carried out in one press stroke is called simple die. Its structure is simple (see Fig. 1-7), so it can be easily manufactured. It is applicable to small batch production.

(2) Progressive Die

The die that several blanking processes are carried out at different positions of the die in one press stroke is called progressive die, as shown in Fig. 1-8. In the operation, the locating pin 2 aims at the locating holes punched previously, and the punch moves downwards to punch by punch 4 and to blank by punch 1, thus the workpiece 8 is produced. When the punch returns, the stripper 6 scrapes the blank 7 from the punch 4, the blank 7 moves forward one step and then the second blanking begins. Above steps are repeated continually. The step distance of the blank is controlled by a stop pin.

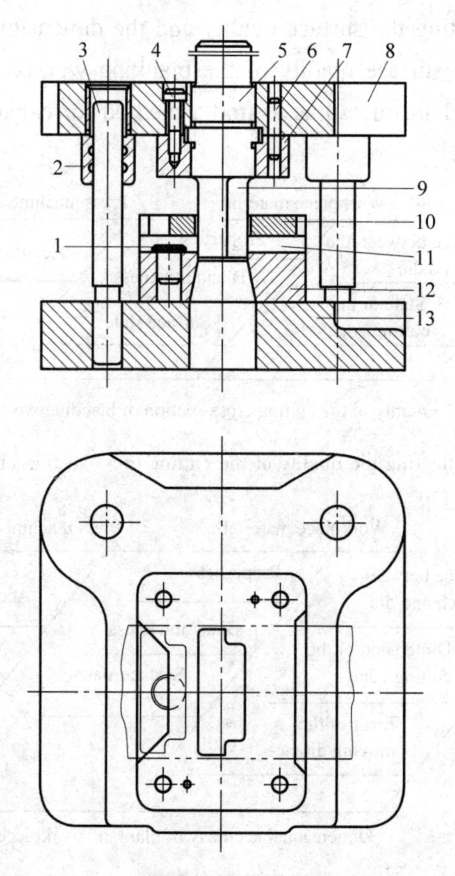

Fig. 1-7 Simple die

1—stop pin; 2—guide bushing; 3—guide pin; 4—bolt; 5—shank; 6—pin; 7—fixed plate; 8—upper bolster; 9—punch; 10—stripper; 11—stock guide; 12—die; 13—lower bolster

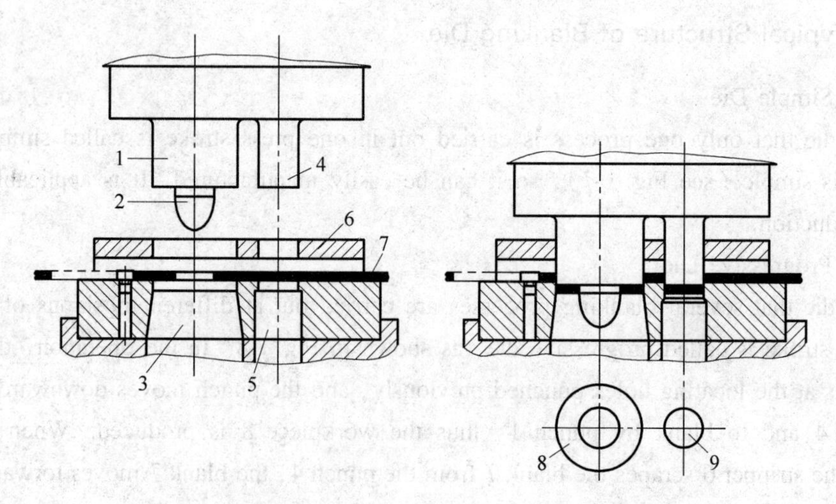

Fig. 1-8 Progressive die for blanking and punching

1—blanking punch; 2—pilot; 3—blanking die; 4—punching punch; 5—punching die; 6—stripper; 7—blank; 8—workpiece; 9—waster

(3) Compound Die

The die that several processes are carried out at the same die position in one press stroke is called compound die, as shown in Fig. 1-9. The main characteristic of the compound die is that the part 1 is both the punch and the die. The outside circle of the punch-die 1 is the cutting edge of the blanking punch, while the inside hole is a deep drawing die. When the slide moves downwards along with the punch-die 1, the blanking process is done first by the punch-die 1 and the blanking die 4, the blanked workpiece is pushed by deep drawing punch 2, and then the deep drawing die moves downwards to carry out deep drawing operation. The ejector 5 and the stripper 3 push the deep drawn workpiece 9 out of the die when the slide returns. The compound die is suitable for mass production and high accuracy blanking.

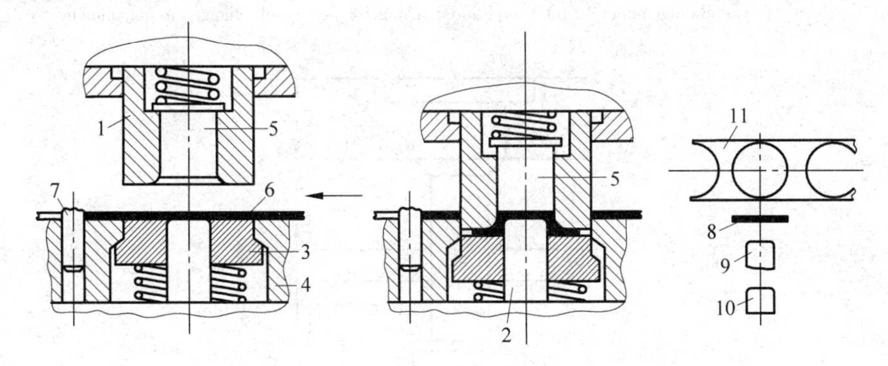

Fig. 1-9 Compound die for blanking and deep drawing

1—punch die; 2—deep drawing punch; 3—press plate (stripper); 4—blanking die; 5—ejector; 6—strip blank; 7—stop pin; 8—blank; 9—deep drawn workpiece; 10—finished part; 11—waste

2. Punch and Die

(1) Punch

There are three kinds of standard punches with circular form in the National Standard, as shown in Fig. 1-10. Which kind of punch should be selected is determined by the dimension d in the working portion. A type circular punch is adopted for $d = 1.1 \sim 30.2$ mm, B type for $d = 3.0 \sim 30.2$ mm, quick-change circular punch for $d = 5 \sim 29$ mm. To fix the circular punch on the punch plate, the hole-base transition fit h6 is adopted for A and B types of circular punches, and the hole-base clearance fit h6 for the quick-change circular punch.

The length L of punch should be determined by the die structure. When using the fixed stripper and stock guide (see Fig. 1-11), the length L of punch is:

$$L = H_1 + H_2 + H_3 + H$$

where, H_1 is the thickness of the fastening plate, in mm; H_2 is the thickness of the stripper, in mm; H_3 is the thickness of the stock guide, in mm; H is the additional length mainly determined by the depth of punch entering into the die (0.5 ~ 1 mm), the total wearing repairing amount (10 ~ 15 mm) and the safe distance between the stripper and punch plate when the die is in the shut state (15 ~ 20 mm).

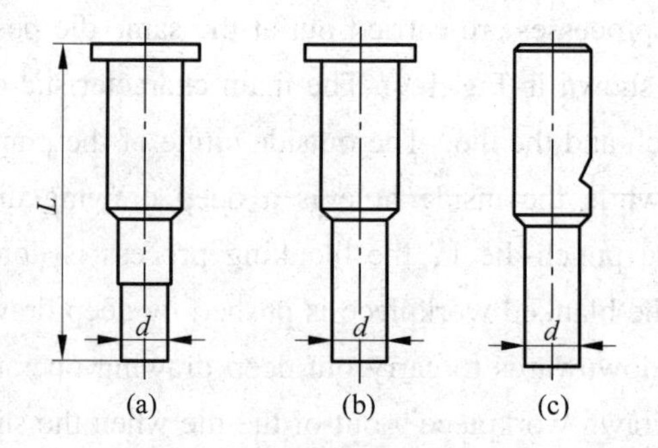

Fig. 1-10 Standard punches

(a) A type circular punch; (b) B type circular punch; (c) quick-change circular punch

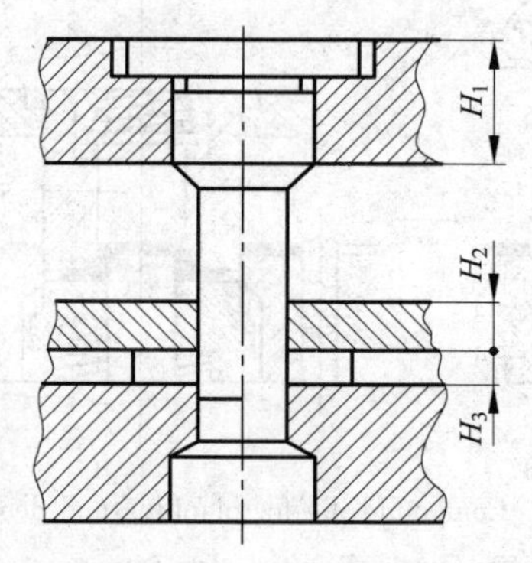

Fig. 1-11 Determination the punch length

The non-standard punches and their fastening patterns are shown in Fig. 1-12. When the distance between punches in the same die is very small, a riveting structure can be used for the circular punch (Fig. 1-12(a)); a jacket structure is usually used for the small hole punching (Fig. 1-12 (b)); the quick-change type is used for the small punch which is vulnerable to damage during blanking (Fig. 1-12(c)); for the non-circular punch, if its size is a bit large, it can be fastened to the die bolster directly by bolts, pins or bolts and location groove instead of

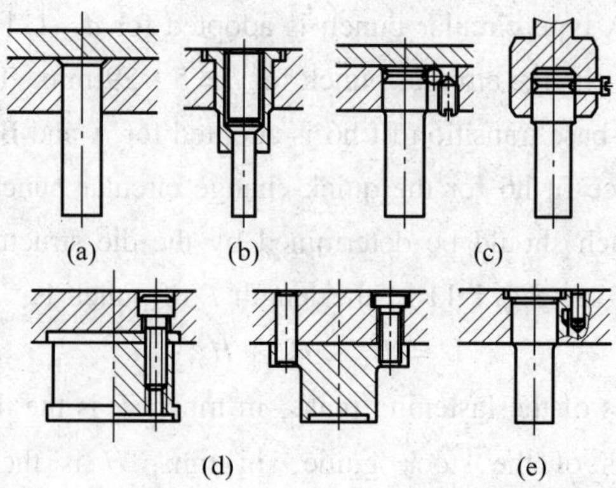

Fig. 1-12 Non-standard punches and their patterns

the fastening plate (Fig. 1-12(d)); if the working portion of punch is non-circular, a circular step structure is used in the fastening portion, and a stop gauge should be added (Fig. 1-12(e)).

For the small punch, it can also be fastened to the punch plate by low-melting alloy, inorganic or epoxy resin adhesive, as shown in Fig. 1-13.

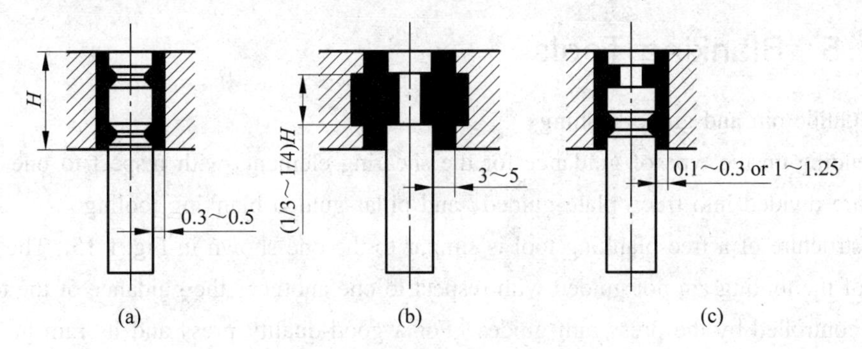

Fig. 1-13 Other methods to fix the punch

(a) by epoxy resin; (b) by low melting point alloy; (c) by inorganic adhesive

(2) Die

The patterns of the die cutting edge are illustrated in Fig. 1-14. Fig. 1-14(a) and Fig. 1-14(b) are the dies with the straight wall cutting edge. The strength of the cutting edge is high, the dimension of its working portion keeps unchanged after mending and its manufacture is convenient. It is suitable for stamping the workpieces with complex shape or high tolerance demand. But the waste of the workpiece in such circumstances is prone to be accumulated inside the hole of the cutting edge, so increasing the expanding force, the ejecting force and the wearing of hole wall. The worn cutting edge forms the shape of inverse cone, which may induce the workpiece jumping from the opening-mouth of the hole to the surface of the die, and cause difficulty in operation. The Fig. 1-14(a) type cutting edge of die is suitable for the non-circular workpiece, and Fig. 1-14(b) type is suitable for the circular workpiece, the die which the workpiece or waste needs to be ejected, or the compound blanking die.

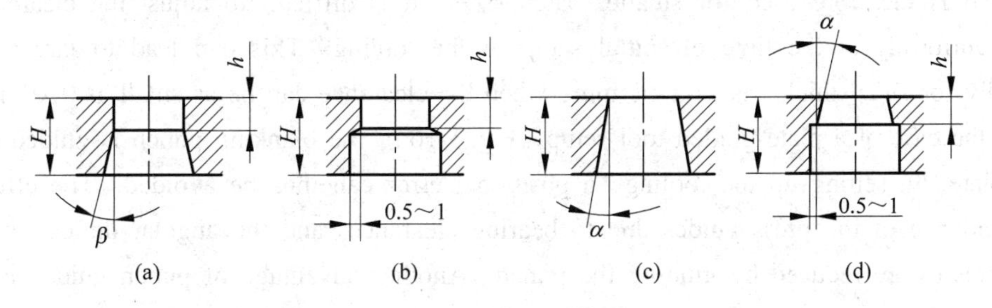

Fig. 1-14 Shapes of the die cutting edge

Fig. 1-14 (c) and Fig. 1-14 (d) show the die with conical shape cutting edge. The workpiece or waste is easy to fall down from the die hole. The workpiece or waste wouldn't accumulate easily inside the hole of the cutting edge. The friction and expanding force exerting on the hole wall are small, therefore the wearing of the die as well as the mending amount of the die per operation are small. But the strength of cutting edge is a bit lower. The dimension of cutting edge increases after mending, but in general its influence on the workpiece dimension and the die life is weak. The dies with conical shape cutting edges are suitable for stamping thin workpieces with simple shape and low tolerance demand.

1.2.5 Blanking Tools

(1) Guide pin and guide bushing

Depending on the type of guidance for the shearing elements with respect to one another, the tools are divided into free, plate-guided, and pillar-guided blanking tooling.

The structure of a free blanking tool is similar to the one shown in Fig. 1-15. The shearing elements of the tooling are not guided with respect to one another; the guidance of the tooling is generally controlled by the press ram guides. For a good-quality press and its ram guides, the tools can be expected to be well guided under load.

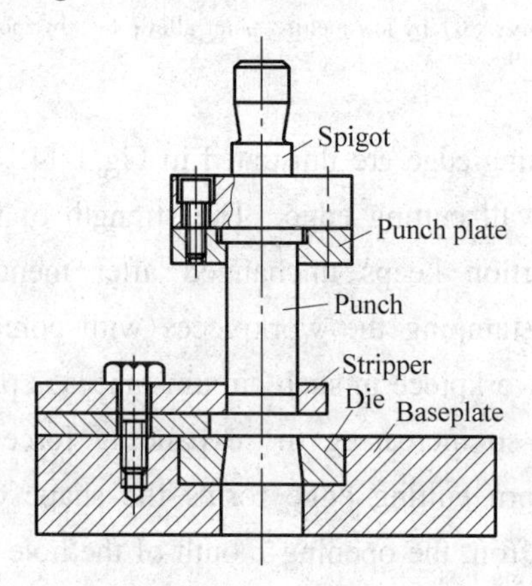

Fig. 1-15 Free blanking tool setup

The free blanking tool is the cheapest type of blanking tooling based on the simplicity of design. It is therefore used for smaller batch sizes. It is difficult to adjust the clearance all around uniformly at the time of initial setup of the tooling. This can lead to larger wear, especially for small thickness as $s \leqslant 1$ mm, when the clearance can be as small as 0.01 mm.

In the case of a plate-guided tool setup (Fig. 1-16), the blanking punch is guided by the guide plate. In setting up the tooling, a positional error can thus be avoided. The effects of poor guidance of the press guides due to bearing clearances and the angular deflection of C-frame presses are reduced by guiding the punch. Another advantage of punch guidance is the resistance of long punches to buckling. The guide plate is also used as a stripper.

The use of a blanking element as a guiding element may pose some disadvantages. If proper measures are not taken, the material particles sticking to the punch or to the punch shoulder will cause a faster wear of the guide plate. Also, the manufacture of large accurate guidance holes for large complicated tooling is difficult and expensive.

In the case of pillar-guided blanking tooling (Fig. 1-17) the functions of guidance and shearing are separated from each other. The pillar guides make this tooling accurate, with the corresponding design of the other elements reducing the tool wear. The errors in bearing clearances of the press need not be taken care of during setting up because of the excellent accuracy of guidance by the pillar die set. Setting the tooling is simple and less time-consuming.

The pillar-guided tooling, like the plate-guided tooling, can contribute to the reduction of defects due to bearing clearance and angular deflection of the press on load. Basically the pillar guides are to be viewed as tools to help in setting up and manufacturing accurate tools. Normally used pillar guides are not stiff enough to take the strong side thrusts caused by noncentral loading on the tools and the tilting movement' in G-type presses without allowing large displacements. Hence the pillar guides are not substitutes for inaccurate press guides and less rigid presses.

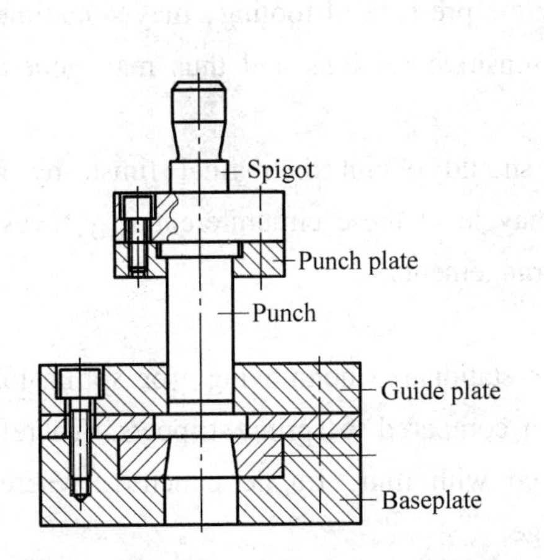

Fig. 1-16 Plate-guided blanking tool setup

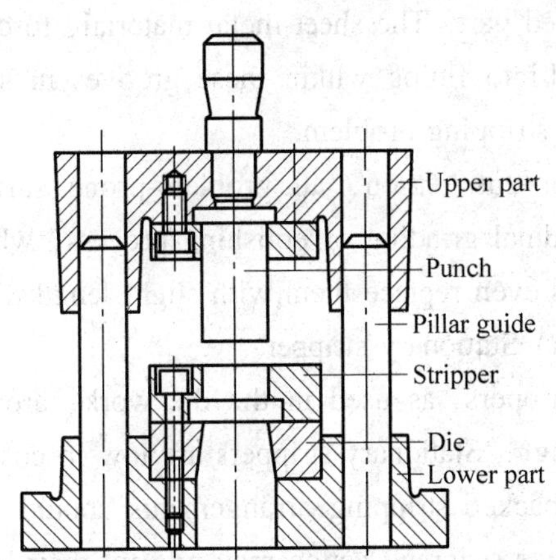

Fig. 1-17 Pillar-guided blanking tool setup

There are different types of pillar guides. Guidance can be accomplished with either bushings or ball bearings (Fig. 1-18). Guides with ball bearings are very rigid under load. They have little friction, and are hence used in fast-stroking presses or in cases where sufficient lubrication is not possible.

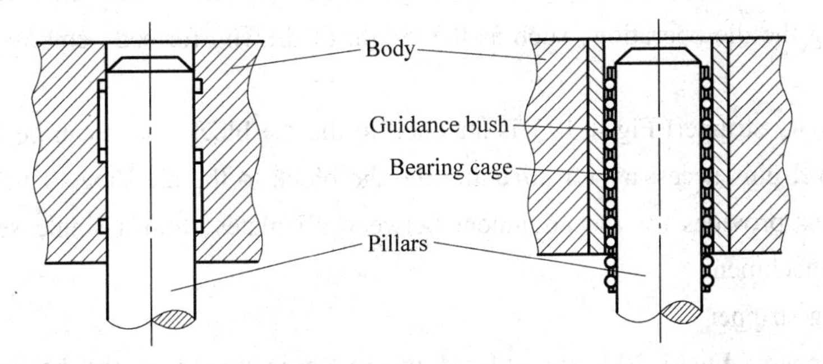

Fig. 1-18 Types of guidance for pillar guides

Pillar-guided tooling can also be equipped with movable guidance plates for the punch.

The guidance plates are mounted on the pillar-guided tooling body and are in general supported by springs on the upper portion of the tool. This design is generally used for blanking thin sheets to ensure the flatness of blanked sheets. The punch can be guided until it touches the sheet. This calls for very precise manufacture of the guidance hole in the plate since there is double guidance, namely, guidance of the punch in the pillar-guided tooling and in the guidance plate. Double guidance is always costly to manufacture.

(2) Stripper

Stripping of parts off the face of the tooling is a complex problem, influenced by the thickness of material and its type, by the surface finish of the strip, and by the surface condition of the tooling as well.

The stripping of parts is further complicated by the prevailing manufacturing procedures, since all conventional metalworking machinery leaves circumferential grooves in the surface of a machined part. The sheet-metal material, forced by the pressure of tooling, may sometimes be coerced into fitting within these grooves in some sensitive sections and thus may generate a serious stripping problem.

For this reason, all problem-prone surfaces should obtain their final finish by some longitudinal grinding or polishing process, which may level these circumferential grooves and perhaps even replace them with slight lengthwise arrangements.

(a) Stationary stripper

Strippers, as used in the die work, are either stationary (nonmoving) or spring-loaded (moving). Stationary strippers are low in cost when compared to spring strippers. Therefore, spring-backed stripping arrangements should be used with thin, fragile punches, where the immediate stripping action may prevent their breakage.

Stationary strippers are of advantage also where additional flattening or material-retaining function is needed or considered beneficial. Such retaining action is usually utilized in drawing, flanging, or other forming operations.

Stationary strippers are provided with a milled channel made to accommodate the strip material. It retains the strip in a fixed location, preventing it from moving anywhere, up, down, or sideways. Naturally, this type of stripper is not adequate where the height of a part is increased during the die operation, such as the height of drawn, formed, embossed, or flanged parts.

The stationary stripper(Fig. 1-19) is attached to the die block and it can be using the same screws and dowel pins necessary for attaching the die block to the die shoe. This way, a single set of dowel pins provides for the alignment between all plates, and a single set of screws is used for their attachment.

(b) Spring stripper

Spring strippers (Fig. 1-20) are utilized where an increased in the height of a part is encountered. They also provide for much firmer stripping action, while acting partially as spring pads during the cutting, forming, or drawing activity of the die.

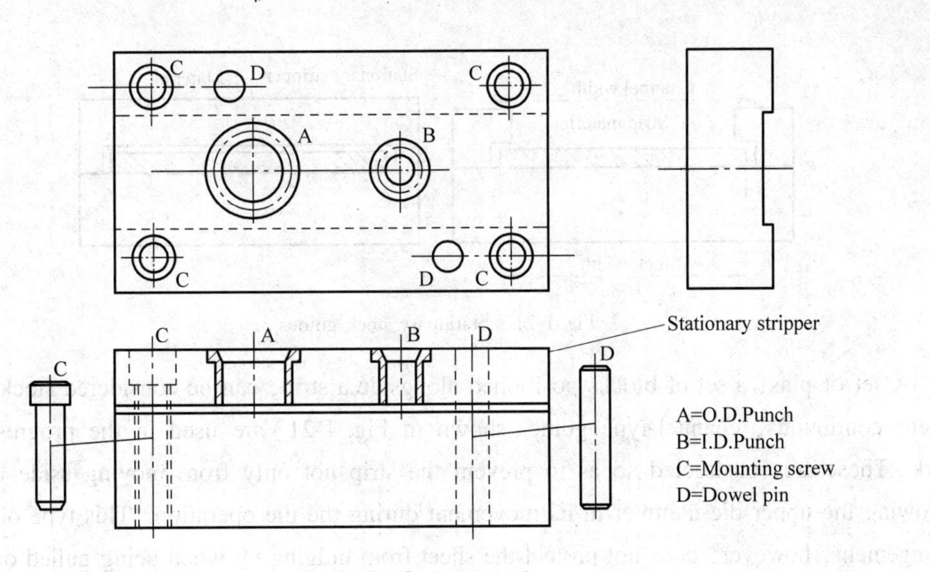

Fig. 1-19 Stationary stripper

Spring strippers are attached to the punch plate, which makes them slide along with the moving of the ram.

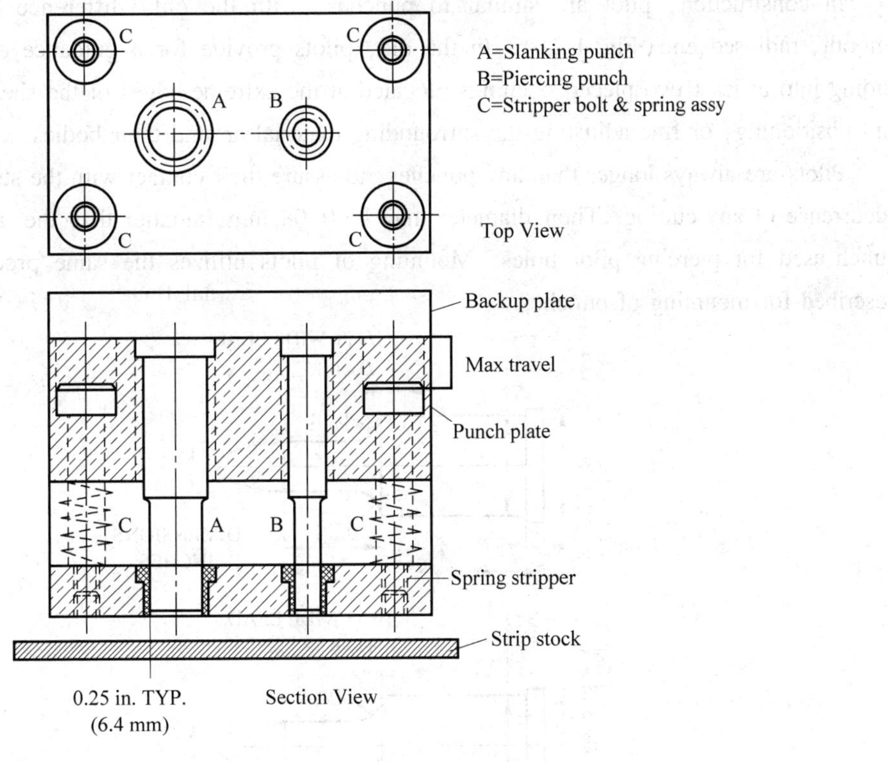

Fig. 1-20 Spring stripper

(3) Stock guide

To guide a strip across the face of a die, stock guides are utilized. These devices can be of various forms and shapes.

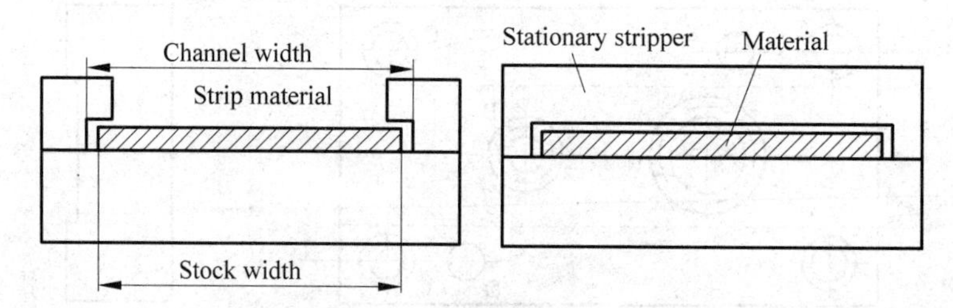

Fig. 1-21 Stationary stock guides

A set of pins, a set of blocks positioned alongside a strip, can be considered stock guides. Often, continuous, channel-type guide (shown in Fig. 1-21) are used in the progressive die work. These are constructed so as to prevent the strip not only from moving aside but from following the upper die member in its movement during the die operation. This type of guiding arrangement, however, does not protect the sheet from bulging up when being pulled on by any of the punches. Such protection can be provided by substituting the side rails with a stationary stripper.

(4) Pilot

In construction, pilot are similar to punches, with the only difference being in their smooth, radiused end (Fig. 1-22). In the die, pilots provide for a guidance of the strip by sliding into at least two pieced openings, located at the extreme edges of the sheet-metal strip, and positioning, or fine-adjusting the surrounding material around their bodies.

Pilots are always longer than any punches, to assure their contact with the strip prior to the occurrence of any cutting. Their diameter may be 0. 08 mm, smaller than the diameter of the punch used for piercing pilot holes. Mounting of pilots utilizes the same procedure as that described for mounting of punches.

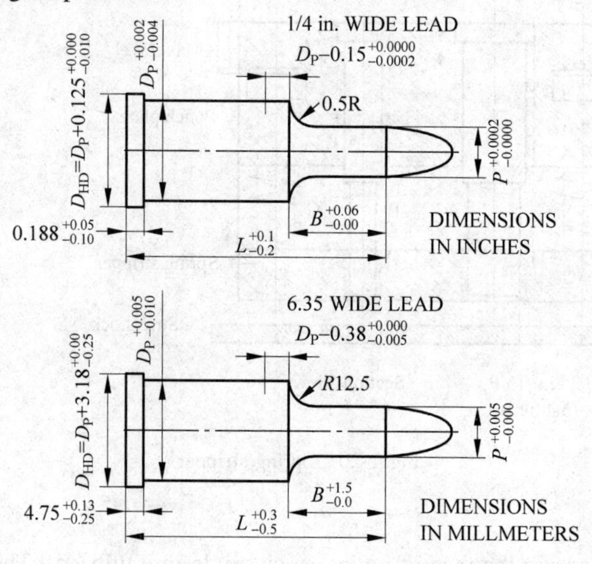

Fig. 1-22 Detail of a pilot

New Words and Expressions

punching 冲孔,冲压
trimming 切边,修边,修整
semi-finished 半制成的
cushion ring 垫片
straightness 直线度
burr 毛刺,飞边
clearance 间隙
ductile 易延展的,柔软的
ductility 延展性
susceptible 敏感的
detrimental effect 不利影响
counterpart 对应的
tolerance 公差
imperfection 缺点,瑕疵
worn-out tools 破旧的工具
force 力
elastic deformation 弹性变形
compression 压缩
tensile 拉伸的
unloading 卸料,卸载
yield strength 屈服强度,流动极限
plastic flow 塑性流动,塑性滑移
sliding deformation 滑移变形
guide plate 导板
tension 拉伸,拉力,张力
simultaneously 同时的,同时发生的
circular angle 圆角
indentation 压入,压痕
squeeze 挤压,压印
crack 裂缝,破裂
equilibrium 平衡,均衡
acting force 作用力
perpendicular to 垂直于
lateral 横向的,侧面的
bearing clearance 轴承间隙
friction 摩擦,摩擦力
pillar die sct 导柱模架
electronic microscope 电子显微镜
dimensional accuracy 尺寸精度
ball bearing 球轴承
fracture zone 断裂区
extrusion 挤压
rollover 塌角
mechanism 机构,机理,机制
simple die 简单模
progressive die 连续模
locating pin 定位销
stripper 脱模杆,卸料板
scrap 废料,切屑
stop pin 定位销,挡料销
slide 滑块
compound die 复合模
ejector 推杆,顶出器
mass production 大量生产
working portion of die 模具工作区
quick-change die 快速凹模,快速换模
press ram 压力机滑块
simplicity 简单,简易
fixed-stripper 固定式卸料板,刚性卸料板
stock guide 导料板,导料装置
fastening plate 固定板,连接板
wearing 磨损
shut 折叠
jacket 罩,套,壳,盖
bolt 螺栓
groove 槽,模膛
gauge 定位装置
low-melting 易熔的,低熔点的
alloy 合金
inorganic 无机的
epoxy resin 环氧树脂
cutting edge 剪刃,刀刃,刃口
ejecting force 顶出力
bushing 衬套,导套

buckle 纵弯,皱纹
noncentral loading 偏心载荷
lubrication 润滑
die life 模具寿命
guide pin 导针,导柱
guide bushing 导套
spigot 塞子,栓
pillar guide 导柱
bearing cage 轴承座
metalworking machinery 金属加工机械
circumferential 圆周的
coerce 强制,迫使
longitudinal 长的,纵的
grinding 磨,碾
polishing 抛光
lengthwise 纵向的
stationary stripper 固定卸料板
spring-loaded 装弹簧的
spring stripper 弹性卸料板
breakage 破损
accommodate 能容纳,可搭载
emboss 凸起
flange 凸缘,法兰
dowel pin 定位销
die block 模板
die shoe 模座
spring pad 弹簧垫
radiused 圆弧形的
mounting 安装,装配

1.3 Bending Dies

Bending is one of the most common forming operations. We merely have to look at the components in an automobile or an appliance——or at a paper clip or a file cabinet——to appreciate how many parts are shaped by bending. Bending is used not only to form flanges, seams, and corrugations but also to impart stiffness to the part (by increasing its moment of inertia).

The terminology used in bending is shown in Fig. 1-23. In bending, the outer fibers of the material are in tension, while the inner fibers are in compression. Because of the Poisson's ratio, the width of the part (bend length, L) in the outer region is smaller, and in the inner region it is larger, than the original width.

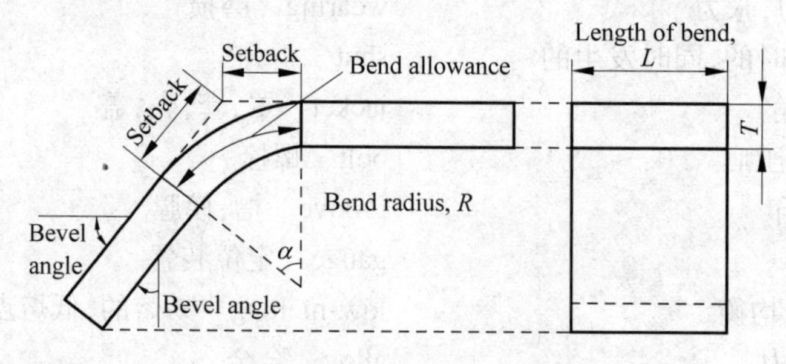

Fig. 1-23 Bending terminology
Note that the bend radius is measured to the inner surface of the bent part.

1.3.1 Simple Bends

When a wire is bent by hand, it takes a natural bend which is governed by the shape of the thumb and finger and the forces applied to the wire. Usually when a bend is made, its radius is governed principally by the tools making the bend.

One leg may be clamped in stationary jaws and a form tool bends the material, as in tube bending, wire bending, pan-brake bending (Fig. 1-24), and dies with clamping pads.

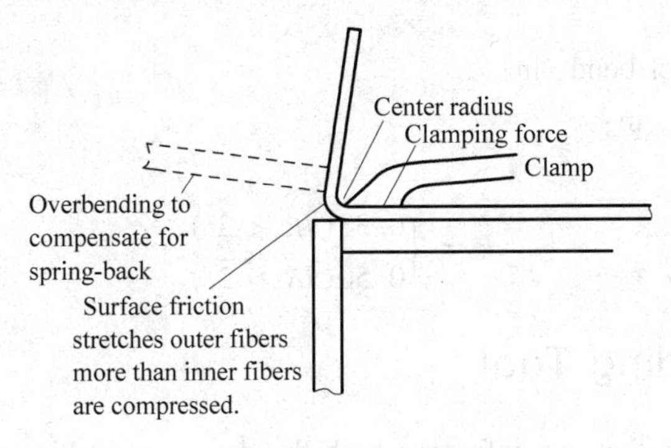

Fig. 1-24 Pan-brake bending

1.3.2 Bending Allowance

The minimum bend radius varies with the alloy and its temper; most annealed sheet metals can be subjected to a bend which has a radius equal to the stock thickness without cracking. The more ductile metals can easily be bent back through a 180° bend, as in a hemming operation.

However, when sheet metal is bent, the total length including the bend is greater than the original stock. This change in length must be considered by the production engineer and die designer because the length of the sheared stock must be known in order to shear the stock.

In the flat position, the neutral axis of a piece of sheet metal coincides with its centerline. But in a bent position, the neutral axis has shifted to a position 0.33 t to 0.40 t from the inner radius (Fig. 1-25). A relationship for determining the length of the developed blank is given below in Fig. 1-25, where the length of the neutral axis is calculated from the circumference of the quad rant of a circle with a radius of $r + t/3$.

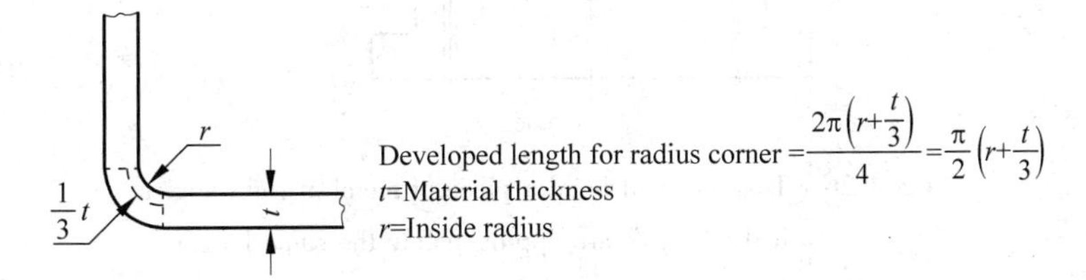

Fig. 1-25 Calculation of the length of blank for a part with a 90°bend

When developing the length of a complex part, first divide it into a series of straight sections, bends, and arcs. Trigonometry can be used to calculate unknown dimensions, but keep the legs of the triangle parallel to the dimension lines, because the hypotenuse is then the bend angle and tile length of the legs can easily be added to or subtracted from the blueprint dimensions.

The length of the bent metal can be calculated from the following empirical relationship:

$$B = \frac{A}{360} 2\pi (IR + kt)$$

where B = bend allowance, in (along the neutral axis);

A = bend angle;

IR = inside radius of bend, in;

t = metal thickness, in;

$$k = \begin{cases} 0.33 (IR < 2t) \\ 0.50 (IR > 2t) \end{cases}.$$

1.3.3 Bending Tool

The various factors that can influence both the design and ultimate working of a bending tool are the thickness of the material, any variation in this thickness, its hardness and temper, the position and severity of the bend, the grain direction and the coupling together of perhaps several shapes with each affecting the other, so making it impossible to determine a final result until the tryout stage of a tool is completed.

Fig. 1-26 shows a simple tool set-up requiring a die, punch and pins as a location medium; this type of tool is used when the straight portions or "legs" of a bend are equal. The bending of soft materials seldom requires any degree of adjustment to the vee to counteract springback, but over bending by five degrees is suggested for harder metals. Initially both the punch and die are made to 85° angle and a slight grinding operation will correct the angle after tryout.

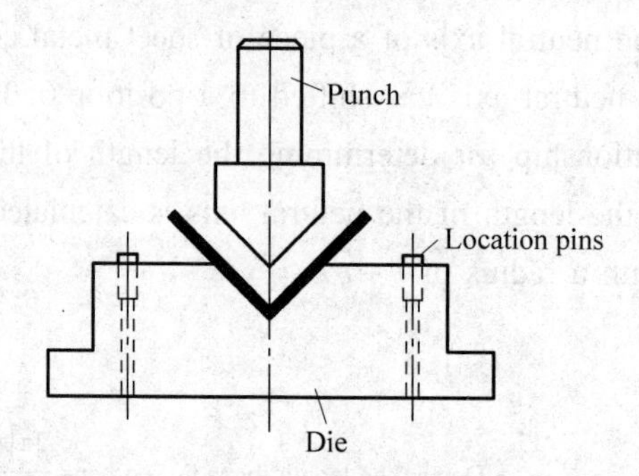

Fig. 1-26 This tool will bend a piece of metal at right angles if the "legs" are approximately the same length

If one of the "legs" takes the shape shown in Fig. 1-27 complications can arise as workpieces will tend to slide as the punch strikes the blank. This prevents the correct length being obtained and although this often is of no consequence, it is a problem that on occasions must receive attention.

The blank must, therefore, be held and Fig. 1-27 shows a workpiece with long "legs" where this is accomplished yet the final tool is not expensive to produce. Tilting the material

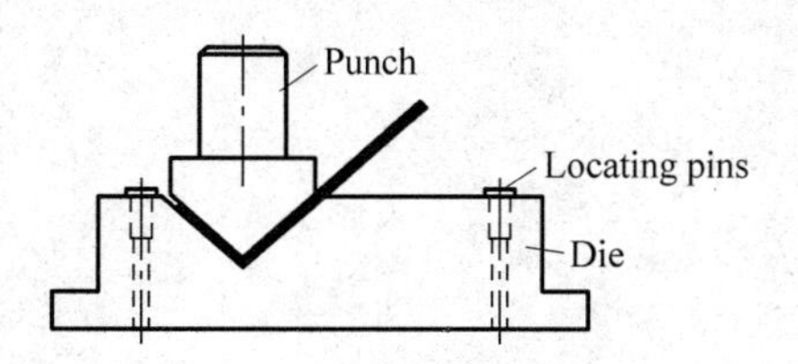

Fig. 1-27 Skidding can occur with this tool when the "legs" differ. However as this is not always important, this design is often provided for such components

means that the adjustment to the vees is possible to ensure that the appropriate angle usually a right-angle and the plunger first holds the blank as the ram descends and recedes into the punch as further movement occurs. Because of the pressure exerted, local hardening of the curved surface is advisable.

New Words and Expressions

automobile 汽车
appliance 设备,工具
file cabinet 文件柜
seam 缝,接缝
corrugation 波纹,皱折,起皱
stiffness 刚性,刚度
inertia 惯性,惯量
terminology 词汇,术语,专门名词
Poisson's ratio 泊松比
clamp 压紧,夹紧
jaw 爪,夹持零件
clamping pad 夹紧垫片
anneal 退火
stock 毛坯,坯料
hemming 折边,卷边
neutral axis 中性轴,中性层
centerline 中心线
circumference 圆周,周边
trigonometry 三角学
hypotenuse (直角三角形的)斜边,弦
subtract 减去,扣除
blueprint 蓝图,设计图
empirical relationship 经验关系
severity 难度
grain direction 晶粒方向
tryout 试验,检验
vee bending tool V 形弯曲模
counteract 抵消,减少
springback 回弹
tilting 倾斜,翻转
plunger 柱塞,滑阀

Plastics Molds

2.1 Introduction

The processing of plastics involves the transformation of a solid (sometimes liquid) polymeric resin, which is in a random form (e.g. powder, pellets, beads), to a solid plastics product of specified shape, dimensions, and properties. This is achieved by means of a transformation process, extrusion, molding, calendering, coating, thermoforming, etc. The process, in order to achieve the above objective, usually involves the following operations: solid transport, compression, heating, melting, mixing, shaping, cooling, solidification, and finishing. Obviously, these operations do not necessarily occur in sequence, and many of them take place simultaneously.

Shaping is required in order to impart the material the desired geometry and dimensions. It involves combinations of viscoelastic deformations and heat transfer, which are generally associated with solidification of the product from the melt.

Shaping includes: (1) two-dimensional operations, e.g. die forming, calendering and coating, and (2) three-dimensional molding and forming operations. Two-dimensional processes are either of the continuous, steady state type (e.g. film and sheet extrusion, wire coating, paper and sheet coating, calendering, fiber spinning, pipe and profile extrusion, etc.) or intermittent as in the case of extrusions associated with intermittent extrusion blow molding. Generally, molding operations are intermittent, and, thus, they tend to involve unsteady state conditions. Thermoforming, vacuum forming, and similar processes may be considered as secondary shaping operations, since they usually involve the reshaping of an already shaped form. In some cases, like blow molding, the process involves primary shaping (parison formation) and secondary shaping (parison inflation).

Shaping operations involve simultaneous or staggered fluid flow and heat transfer. In two-dimensional processes, solidification usually follows the shaping process, whereas solidification and shaping tend to take place simultaneously inside the mold in three dimensional processes. Flow regimes, depending on the nature of the material, the equipment, and the processing conditions, usually involve combinations of shear, extensional, and squeezing flows in conjunction with enclosed (contained) or free surface flows.

The thermo-mechanical history experienced by the polymer during flow and solidification results in the development of microstructure (morphology, crystallite, and orientation distributions) in the manufactured article. The ultimate properties of the article are closely related to the microstructure. Therefore, the control of the process and product quality must be based on an understanding of the interactions between resin properties, equipment design, operating conditions, thermo-mechanical history, microstructure, and ultimate product properties.

New Words and Expressions

polymer 聚合体
polymeric resin 聚合树脂
powder 粉,粉末
pellet 丸,粒
bead 微珠,空心颗粒
transformation 转变,变换,相变
molding 造型,制模
coating 涂层,涂料
thermoforming 热成形
solidification 凝固,固化
finishing 精加工
geometry 几何学,几何形
viscoelasticity 黏弹性
die-forming 模具成形
calendering 压延,压制成形
film 薄膜
crystallite 结晶性(度)
intermittent 间歇的,断续的
blow molding 吹塑模
vacuum forming 真空成形
secondary operation 二次加工
reshaping 整形
parison 型坯
stagger 交错,错移
fluid flow 流体流量
heat transfer 热传导,传热,导热
two-dimensional forming 二维成形
three-dimensional forming 三维成形
pipe extrusion 管材挤压
thermo-mechanical 热-机的
article 项目,产品
microstructure 显微组织,微观结构
morphology 形态学
orientation 位向,取向,定向

2.2 The Properties of Plastics

Plastics are organic materials made from large molecules that are constructed by a chain-like attachment of certain building-block molecules. The properties of the plastic depend heavily on the size of the molecule and on the arrangement of the atoms within the molecule. For example, polyethylene is made from the ethylene building block that is initially a gas. Through a process called polymerization, a chain of ethylene molecules is formed by valence bonding of the carbon atoms within the ethylene molecule. The high molecular weight product which results is called a polymer. Hence, the designation of polyethylene is used to distinguish the high- molecular-weight plastic from its gaseous counterpart, ethylene, which is the monomer that becomes polymerized. The "poly" refers to the "many" ethylene building block molecules or monomers, which join to form the polyethylene plastic molecule. Frequently, the term "resin" is used, interchangeably with "polymer" to describe the backbone molecule of a plastic material. However, "resin" is sometimes used to describe a syrupy liquid of both natural and synthetic resin.

Plastics, in the finished product form, are seldom comprised exclusively of polymer but also include other ingredients such as fillers, pigments, stabilizers, and processing aids. However, designation of the plastic material or molding compound is always taken from the polymer designation.

Broadly speaking plastics may be divided into two categories: thermoplastics and thermoset plastics. The classes of materials are so named because of the effect of temperature on their properties.

2.2.1 Thermosets

Thermoset plastics are polymers which are relatively useless in their raw states. Upon heating to a certain temperature a chemical reaction takes place which causes the molecules to bond together or cross-link. After vulcanization and polymerization, or curing, the thermoset material remains stable and cannot return to its original state. Thus, "thermo-set" identifies those materials that become set in their useable state resulting from the addition of heat. Normally, a thermoset polymer is mixed with fillers and reinforcing agents to obtain the properties of a molding compound.

Thermosets are the hardest and stiffest of all plastics, are chemically insoluble after curing, and their properties are less affected by changes in temperature than the heat-sensitive thermoplastics. The closest non-plastic counterparts to thermosets in properties are ceramics. Common examples of thermoset plastics are: phenolics, melamine, urea, alkyds, and epoxies. Molding compounds made from these polymeric resins always contain additional fillers and reinforcing agents to obtain optimum properties.

2.2.2 Thermoplastics

Thermoplastic polymers are heat-sensitive materials which are solids at room temperature, like most metals. Upon heating, the thermoplastics begin to soften and eventually reach a melting point and become liquid. Allowing a thermoplastic to cool below its melting point causes resolidification or freezing of the plastic. Successive heating and cooling cycles cause repetition of the melting-freezing cycle just as it does for metals.

The fact that thermoplastics melt is the basis for their processing into finished parts. Thermoplastics may be processed by any method which causes softening or melting of the material. Examples of thermoplastic fabrication techniques using melting are: injection molding, extrusion, rotational casting, and calendering. Fabrication methods which take advantage of softening below the melting point are: thermoforming (vacuum or pressure), blow molding, and forging. Of course, normal metal-cutting techniques can also be applied to thermoplastics in the solid state. Common examples of thermoplastics are: polyethylene, polystyrene, polyvinyl chloride (PVC), and nylon (polyamide).

2.2.3 Fillers

Plastics often contain other added materials called fillers. Fillers are employed to increase bulk and to help impart desired properties. Plastics containing fillers will cure faster and hold closer to established finished dimensions, since the plastic shrinkage will be reduced. Wood flour is the general-purpose and most commonly used filler. Cotton frock, produced from cotton linters, increases mechanical strength. For higher strength and resistance to impact, cotton cloth chopped into sections about 1/2-inch square can be processed with the plastic. Asbestos fiber may be used as a filler for increasing heat and fire resistance, and mica is used for molding plastic parts with superior dielectric characteristics. Glass fibers, silicon, cellulose, clay, or nutshell flour may also be used. Nutshell flour is used instead of wood flour where a better finish is desired. Plastic parts using short fiber fillers will result in lower costs, while those with long fiber fillers having greater impact strengths are more expensive. Other materials, not defined as fillers, such as dyes, pigments, lubricants, accelerators, and plasticizers may also be added. Plasticizers are added to soften and improve the moldability of plastics. Filler and modifying agents are added and mixed with the raw plastic before it is molded or formed.

2.2.4 Properties of Plastics

1. General Properties

The problem of selecting plastic materials is that of finding the material with suitable properties from the standpoint of intended service, methods of forming and fabricating, and cost. New and improved plastic materials possessing almost any desired characteristic are being introduced continually. There are plastics that do not require plasticizers that have greater flexibility under lower temperatures, and are stable under higher temperatures. Some resist water, acids, oils, and other destructive matter. The wide use of plastics testifies to their value; however, fundamental limitations should be considered when applying a new material or adapting an old material to new applications.

2. Mechanical Properties

Several unfamiliar aspects of material behavior of plastics need to be appreciated, the most important probably being that, in contrast to metals at room temperature, the properties of plastics are time dependent. Then superimposed on this aspect are the efforts of the level of stress, the temperature of the material, and its structure (such as molecular weight, molecular orientation, and density). For example, with polypropylene an increase in temperature from 20 to 60℃ may typically cause a 50% decrease in the allowable design stress. In addition, for each 0.001 g/cm^3 change in density of this material there is a corresponding 4% change in design stress. The material, moreover, will have enhanced strength in the direction of molecular alignment (that is, in the direction of flow in the mold) and less in the transverse direction.

(1) Stress-Strain Behavior

Any force or load acting on a body results in stress and strain in the body. Stress represents the intensity of the force at any point in the body. The stress-strain behavior of plastics measured at a constant rate of loading provides a basis for quality control and comparative evaluation of various plastics. The diagram shown in Fig. 2-1 is most typical control of that obtained in tension for a constant rate of loading. For compression and shear the behavior is quite similar except that the magnitude and the extent to which the curve is followed are different.

In the diagram, load per unit cross-section (stress) is plotted against deformation expressed as a fraction of the original dimension (strain). Even for different materials the nature of the curves will be similar, but they will differ in (1) the numerical values obtained and (2) how far the course of the typical curve is followed before failure occurs. Cellulose acetate and many other thermoplastics may follow the typical curve for almost its entire course. Thermosets like phenolics, on the other hand, have cross-linked molecules, and only a limited amount of intermolecular slippage can occur. As a result, they undergo fracture at low strains, and the stress-strain curve is followed no further than to some point below the knee, such as point 1.

Ultimate strength, elongation, and elastic modulus (Young's modulus) can be obtained from the stress-strain study. The appearance of a permanent set is said to mark a yield point, which indicates the upper limit of usefulness for any material. Unlike some metals, in particular, the ferrous alloys, an arbitrary yield point is usually assigned to them. Typical of these arbitrary values is the 0.2% or the 1% offset yield stress.

Up to point 1 in Fig. 2-1, the material behaves as an elastic solid, and the deformation is recoverable. This deformation, which is small, is associated with the bending or stretching of the inter-atomic bonds between atoms of the polymer molecules. This type of deformation is nearly instantaneous and recoverable. There is no permanent displacement of the molecules relative to each other.

Between points 1 and 2 in Fig. 2-1, deformations have been associated with a straightening out of a kinked or coiled portion of the molecular chains, if loaded in tension. This can occur without intermolecular slippage. The deformation is recoverable ultimately but not instantaneously and hence is analogous to that of a nonlinear spring. Although the deformation occurs at stresses exceeding the stress at the proportional limit, there is no permanent change in intermolecular arrangement. This kind of deformation, characterized by recoverability and

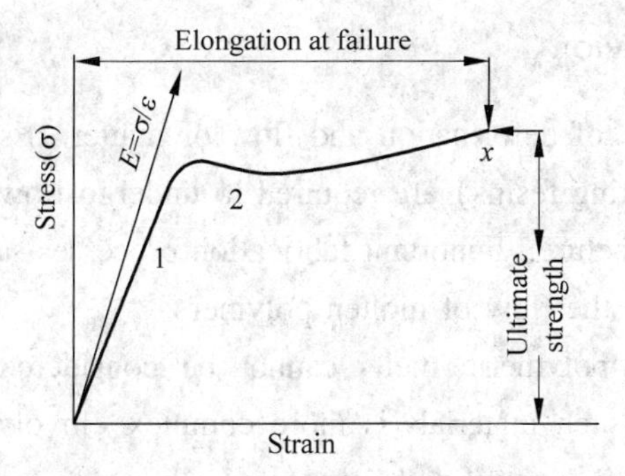

Fig. 2-1 Nominal stress – strain diagram

nonlinearity, is very pronounced in the rubber state.

The greatest extension that is recoverable marks the elastic limit for the material. Beyond this point extensions occur by displacement of molecules with respect to each other, as in Newtonian flow of a liquid. The displaced molecules have no tendency to slip back to their original positions, therefore these deformation are permanent and not recoverable.

(2) Viscoelastic Behavior of Plastics

In a perfectly elastic (Hookean) material the stress, σ, is directly proportional to the strain, ε. For uniaxial stress and strain the relationship may be written as

$$\sigma = \text{constant} \times \varepsilon \tag{2.1}$$

where the constant is referred to as the modulus of elasticity.

In a perfectly viscous (Newtonian) liquid the shear stress, τ, is directly proportional to the rate of strain, $\dot{\gamma}$, and the relationship may be written as

$$\tau = \text{constant} \times \dot{\gamma} \tag{2.2}$$

where the constant is reffered to as the viscosity.

Polymeric materials exhibit stress-strain behavior which falls somewhere between these two ideal cases; hence, they are termed viscoelastic. In a viscoelastic material the stress is a function of both strain and time and so may be described by an equation of form

$$\sigma = f(\varepsilon, t) \tag{2.3}$$

This equation represents nonlinear viscoelastic behavior. For simplicity of analysis it is often reduced to the form

$$\sigma = \varepsilon f(t) \tag{2.4}$$

which represents linear viscoelasticity. It means that in a tensile test on linear viscoelastic material, for a fixed value of elapsed time, the stress will be directly proportional to strain.

The most characteristics features of viscoelastic materials are that they exhibit time-dependent deformation or strain when subjected to a constant stress (creep) and a time-dependent stress when subjected to a constant strain (relaxation). Viscoelastic materials also have the ability to recover when the applied stress is removed. To a first approximation, this recovery can be considered as a reversal of creep.

3. Rheological Behavior

Rheology is the science of deformation and flow of matter. Essentially, all thermoplastic resins (and many thermosetting resins) are required to undergo flow in the molten state during the course of product manufacture. Important fabrication processes such as injection, extrusion, and calendaring, all involve the flow of molten polymers.

The flow behavior of polymeric melts cannot be considered to be purely viscous in character. The response of such materials is more complex, involving characteristics that are both viscous and elastic. This is only to be expected when one is trying to deform variously entangled long-chain molecules with a distribution of molecular weights.

During flow, polymer molecules not only slide past each other, but also tend to uncoil-or at least they are deformed from their equilibrium, random coiled-up configuration. On release of the deforming stresses, these molecules tend to revert to random coiled-up forms. Since molecular entanglements cause the molecules to act in a cooperative manner, some recovery of shape corresponding to the recoiling occurs.

We begin by making a reference to Fig. 2-2, which schematically illustrates two parallel plates of very large area A separated by a distance r with the space in between filled with a liquid. The lower plate is fixed and a shear force F_S is applied to the top plate of area A producing a shear stress($\tau = F_S/A$) that causes the plate to move at a uniform velocity v in a direction parallel to the direction of the force. It may be assumed that the liquid wets the plates and that the molecular layer of liquid adjacent to the stationary plate is stationary while the layer adjacent to the top plate moves at the same velocity gradient between the two plates is $\mathrm{d}v/\mathrm{d}r$. It is defined as the *shear rate* and is commonly given the symbol $\dot{\gamma}$ i. e. ,

$$\dot{\gamma} = \mathrm{d}v/\mathrm{d}r \tag{2.5}$$

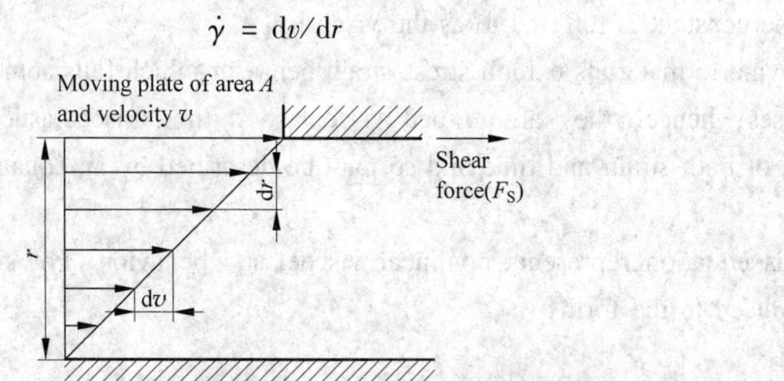

Fig. 2-2 Velocity distribution of a liquid between two parallel plates, one stationary and the other moving

If the liquid is ideal and it is maintained at a constant temperature, the shear stress is linearly and directly proportional to the shear rate such that one way write

$$\tau = \eta(\mathrm{d}v/\mathrm{d}r) = \eta\dot{\gamma} \quad \text{or} \tag{2.6}$$

$$\eta = \tau/\dot{\gamma} \tag{2.7}$$

where η is the coefficient of viscosity or simply the viscosity or internal friction of the liquid.

The linear relationship between τ and $\dot{\gamma}$ given by Equ. (2.6), or Equ. (2.7) is known as Newton's law and liquids which behave in this manner are called Newtonian fluids or ideal fluids. Other fluids which deviate from Newton's law are described as non-Newtonian. For such fluids, the viscosity defined by Equ. (2.7) is also known as the apparent viscosity.

In practice, the Newtonian behavior is confined to low molecular weight liquids. Polymer melts obey Newton's law only at shear rates close to zero and polymer solution only at concentrations close to zero. The most general rheological equation is

$$\tau = K\dot{\gamma}^{n} \tag{2.8}$$

Several common types of rheological behavior are shown in Fig. 2-3. These flow phenotypes are named pseudoplastic, dilatant, and Bingham. In Newtonian liquids, the viscosity is constant and independent of shear rate.

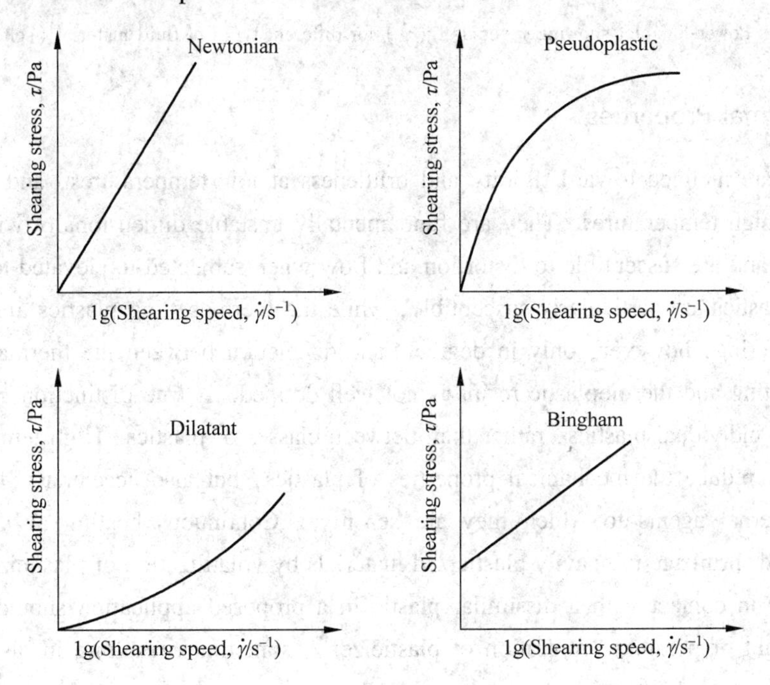

Fig. 2-3 Flow curves (τ versus $\dot{\gamma}$) for different types of fluid material

In pseudoplastic and dilatant liquids the viscosity is no longer constant. In the former it decreases and in the latter it increases with increasing shear rate; that is to say, the shear stress increases with increasing shear rate less than proportionately in a pseudoplastic and more than protionately in a dilatant. Pseudoplastics are thus described as shear-thinning and dilatants as shear-thickening fluid systems.

Where K is the zero shear (Newtonian) viscosity. Equ. (2.8) gives a linear relationship between lg τ and lg $\dot{\gamma}$ and the slope of the experimental plot (Fig. 2-4) gives the value of n. Polymer melts are almost of the pseudoplastic type.

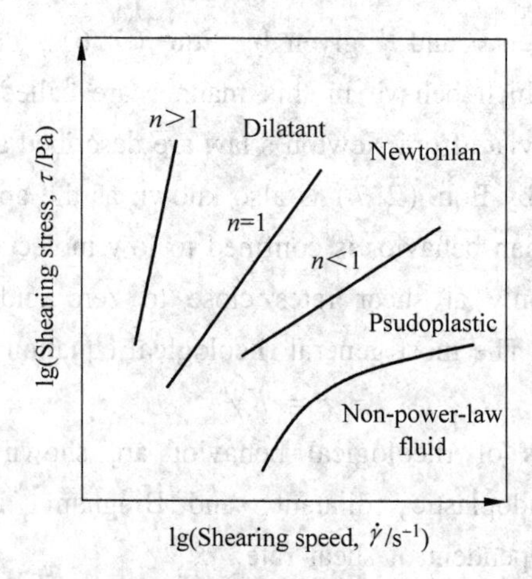

Fig. 2-4 Power-low plot showing τ versus lg($\dot{\gamma}$) for different types of fluid material (schematic)

4. Thermal Properties

Plastics are inclined toward rigidity and brittleness at low temperatures, and softness and flexibility at high temperatures. They are fundamentally unstable dimensionally with respect to temperature, and are susceptible to distortion and flow when subjected to elevated temperatures. The thermoplastics are particularly susceptible, while the thermosetting plastics are much more resistant, differing, however, only in degree. The distinction between the thermal stability of the thermosetting and thermoplastic resins is not well defined. A true distinction can be drawn only between individual plastics, rather than between classes of plastics. High temperatures not only seriously reduce the mechanical properties of plastics, but also accelerate the destructive action of external agents to which they are sensitive. Continuous heating also may induce brittleness and shrinkage in heavily plasticized materials by volatilization of plasticizers. The use of one plastic in contact with a dissimilar plastic in a proposed application should be checked first in the light of possible "migration of plasticizer", sometimes resulting in discoloration or hardening of one of the plastics.

In general, moderate temperatures are required for storage of plastics over long periods; low temperatures are to be avoided because of the low-temperature brittleness of most of the plastics, and high temperatures should be avoided because of the rapid loss of mechanical properties, volatilization of plasticizers, and the susceptibility of a large number to distortion.

5. Optical Properties

Prolonged exposure to sunlight will affect adversely all plastics with exception of tetrafluoroethylene (Teflon). The change induced by the ultraviolet components may vary in kind and severity from slight yellowing to complete disintegration as a result of the chemical degradation of the polymeric compound or plasticizers. Loss of strength, reduced ductility, and increased fragility usually accompany such action. Many plastics are offered in special formulations containing "ultraviolet inhibitors" which should be utilized when this influence is present. Exposure of plastics to sunlight during storage should be avoided, especially when the transparency of clear materials is to be preserved.

6. Electrical Properties

Plastics have excellent electrical resistivity, which makes them having wide application as an insulating material. In the high-frequency applications, plastics are particularly advantageous and, consequently, are being used to a large extent in the fields of radar and television.

7. Reinforced Plastics

The modulus and strength of plastics can be increased significantly by means of reinforcement. A reinforced plastic consists of two main components—a matrix, which may be either a thermoplastic or thermosetting resin, and a reinforcing filler, which is mostly used in the form of fibers (but particles, for example glass spheres, are also used.)

The greater tensile and stiffness of fibers as compared with the polymer matrix is utilized in producing such composites. In general, the fibers are the load-carrying members, and the main role of the matrix is to transmit the load to the fibers, to protect their surface, and to raise the energy for crack propagation, thus preventing a brittle-type fracture. The strength of the fiber-reinforced plastics is determined by the strength of the fiber and by the nature and strength of the bond between the fibers and the matrix.

New Words and Expressions

organic 有机的
molecule 分子
attachment 附件,配件,装配装置
building block 标准部件,结构单元
polyethylene 聚乙烯
ethylene 乙烯
polymerization 聚合
valence 化合价,原子价
carbon atoms 碳原子
designation 命名,指示
high-molecular-weight 高分子量
gaseous 气体的,气态的
monomer 单体
resin 树脂,涂树脂于
interchangeably 可交换地,可替换地
backbone 中枢,支柱
syrupy 像糖浆的,糖浆似的
syrupy liquid 黏流体
synthetic 合成的
ingredient 成分,因素
filler 装填物,漏斗,衬垫,焊补料层
pigment 颜料
stabilizer 稳定剂
processing aid 加工助剂
category 种类,类别
thermoplastics 热塑性塑料
thermoset plastics 热固性塑料
chemical reaction 化学反应
bond 黏合
cross-link 交叉结合
vulcanization 硫化,硫化过程
curing 硫化,固化
reinforcing 强化,加强
molding compound 模塑料

insoluble 不能溶解的
heat-sensitive 热敏的
non-plastic 非塑性的
phenolic 酚醛树脂
melamine 三聚氰胺
urea 尿素
alkyd 醇酸树脂
melting point 熔点
resolidification 再凝固作用
repetition 重复,循环
fabrication 制作,构成,伪造物,装配工
injection 注射
rotational casting 离心浇铸
blow molding 吹塑成形
forging 锻造
metal-cutting technique 金属切削技术
polystyrene 聚苯乙烯
polyvinyl chloride 聚氯乙烯
nylon 尼龙
polyamide 聚酰胺
shrinkage 收缩
wood flour 木粉
cotton linter 棉绒
cotton cloth 棉布
chop 砍,剁碎
asbestos 石棉
mica 云母
dielectric 绝缘体,非传导性的
glass fiber 玻璃纤维
silicon 硅
cellulose 纤维素
clay 黏土,泥土
nutshell flour 坚果壳粉
short fiber filler 短纤维填充剂
long fiber filler 长纤维填充剂
dye 染,染色
lubricant 润滑剂
accelerator 加速器
plasticizer 可塑剂
moldability 可塑性
standpoint 立场,观点
flexibility 弹性,机动性,挠性
superimposed 叠加
polypropylene 聚丙烯
stress-strain 应力应变
magnitude 大小,数量,巨大
intensity 强度
constant rate 恒定速度
curve 曲线
cellulose acetate 醋酸纤维素
cross-linked molecule 交联的分子
intermolecular slippage 分子间滑动
elastic modulus 弹性模量
Young's modulus 杨氏模量
yield point 屈服点
ferrous 有色金属
arbitrary 任意,随意
elastic solid 弹性体
kink 扭结,弯曲
coiled 盘绕
slippage 滑动
analogous 类似于
viscoelastic behavior 黏弹行为
Hookean 胡克
uniaxial 单轴,单向
elasticity 弹性,弹力
viscosity 黏度
constant stress 恒定应力
relaxation 缓和,放松
rheological 流变的,流变学的
rheology 流变学
molten state 熔融状态
entangle 卷入,纠缠在一起
schematically 图表
coefficient 系数
apparent viscosity 表观黏度
phenotype 显型,表现型
dilatant 膨胀剂

Bingham 宾汉姆
shear-thinning 剪切稀化
brittleness 脆性
softness 柔性,柔和
with respect to 关于,至于
distortion 扭曲,变形
thermal stability 热稳定性
volatilization 挥发
discoloration 变色,污点
tetrafluoroethylene(Teflon) 聚四氟乙烯
ultraviolet 紫外线
disintegration 解体
ultraviolet inhibitor 紫外线抑制剂
transparency 透明度
susceptibility 敏感性,磁化系数
formulations 配方
storage 存储
high-frequency 高频
matrix 母体,基础
propagation 扩展
brittle-type 脆性断裂

2.3 Injection Molds

2.3.1 Injection Molding

Injection molding is principally used for the production of thermoplastic parts, and it is also one of the oldest. Currently injection-molding accounts for 30% of all plastics resin consumption. Typical injection-molded products are cups, containers, housings, tool handles, knobs, electrical and communication components (such as telephone receivers), toys, and plumbing fittings.

Polymer melts have very high viscosities due to their high molecular weights; they cannot be poured directly into a mold under gravity flow as metals can, but must be forced into the mold under high pressure. Therefore while the mechanical properties of a metal casting are predominantly determined by the rate of heat transfer from the mold walls, which determines the grain size and grain orientation in the final casting, in injection molding the high pressure during the injection of the melt produces shear forces that are the primary cause of the final molecular orientation in the material. The mechanical properties of the finished product are therefore affected by both the injection conditions and the cooling conditions within the mold.

Injection molding has been applied to thermoplastics and thermosets, foamed parts, and has been modified to yield the reaction injection molding (RIM) process, in which the two components of a thermosetting resin system are simultaneously injected and polymerize rapidly within the mold. Most injection molding is however performed on thermoplastics, and the discussion that follows concentrates on such moldings.

A typical injection molding cycle or sequence consists of five phases (see Fig. 2-5):

(1) Injection or mold filling;

(2) Packing or compression;

(3) Holding;

(4) Cooling;

(5) Part ejection.

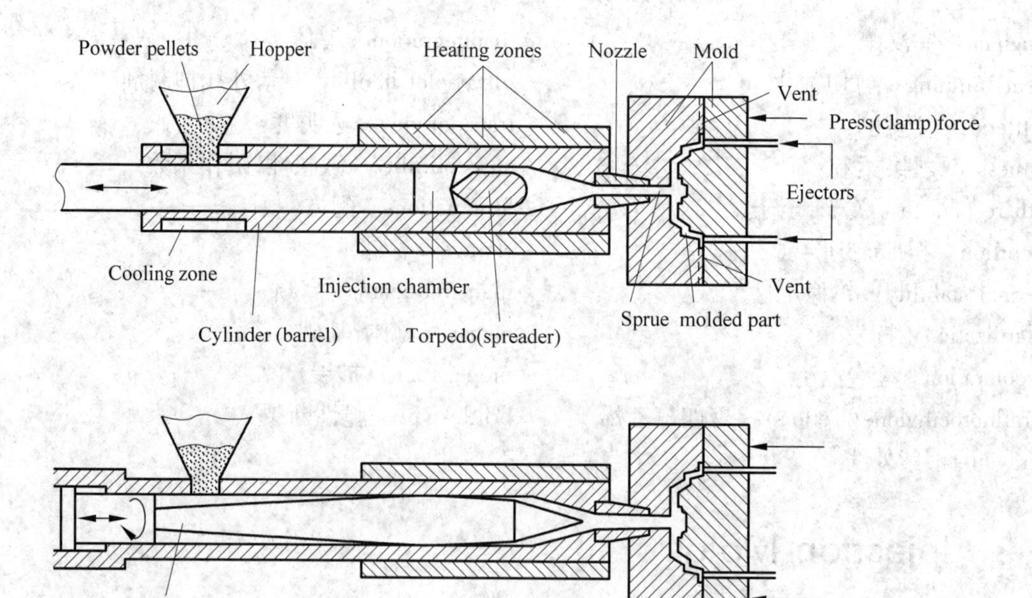

Fig. 2-5 Injection molding process

Plastic pellets (or powder) are loaded into the feed hopper and through an opening in the injection cylinder where they are carried forward by the rotating screw. The rotation of the screw forces the pellets under high pressure against the heated walls of the cylinder causing them to melt. Heating temperatures range from 265 to 500 °F. As the pressure builds up, the rotating screw is forced backward until enough plastic has accumulated to make the shot. The injection ram (or screw) forces molten plastic from the barrel, through the nozzle, sprue and runner system, and finally into the mold cavities. During injection, the mold cavity is filled volumetrically. When the plastic contacts the cold mold surfaces, it solidifies (freezes) rapidly to produce the skin layer. Since the core remains in the molten state, plastic flows through the core to complete mold filling. Typically, the cavity is filled to 95% ~98% during injection.

Then the molding process is switched over to the packing phase. Even as the cavity is filled, the molten plastic begins to cool. Since the cooling plastic contracts or shrinks, it gives rise to defects such as sink marks, voids, and dimensional instabilities. To compensate for shrinkage, addition plastic is forced into the cavity. Once the cavity is packed, pressure applied to the melt prevents molten plastic inside the cavity from back flowing out through the gate. The pressure must be applied until the gate solidifies. The process can be divided into two steps (packing and holding) or may be encompassed in one step (holding or second stage). During packing, melt forced into the cavity by the packing pressure compensates for shrinkage. With holding, the pressure merely prevents back flow of the polymer melt.

After the holding stage is completed, the cooling phase starts. During cooling, the part is held in the mold for specified period. The duration of the cooling phase depends primarily on the material properties and the part thickness. Typically, the part temperature must cool below the material's ejection temperature.

While cooling the part, the machine plasticates melt for the next cycle. The polymer is subjected to shearing action as well as the condition of the energy from the heater bands. Once the shot is made, plastication ceases. This should occur immediately before the end of the cooling phase. Then the mold opens and the part is ejected.

2.3.2 Injection Molds

Molds for injection molding are as varied in design, degree of complexity, and size as the parts produced from them. The functions of a mold for thermoplastics are basically to impart the desired shape to the plasticized polymer and then to cool the molded part.

A mold is made up of two sets of components: (1) the cavities and cores, and (2) the base in which the cavities and cores are mounted. The size and weight of the molded parts limit the number of cavities in the mold and also determine the equipment capacity required. From consideration of the molding process, a mold has to be designed to safely absorb the forces of clamping, injection, and ejection. Also, the design of the gates and runners must allow for efficient flow and uniform filling of the mold cavities.

Fig. 2-6 illustrates the parts in a typical injection mold. The mold basically consists of two parts: a stationary half (cavity plate), on the side where molten polymer is injected, and a moving half (core plate) on the closing or ejector side of the injection molding equipment. The

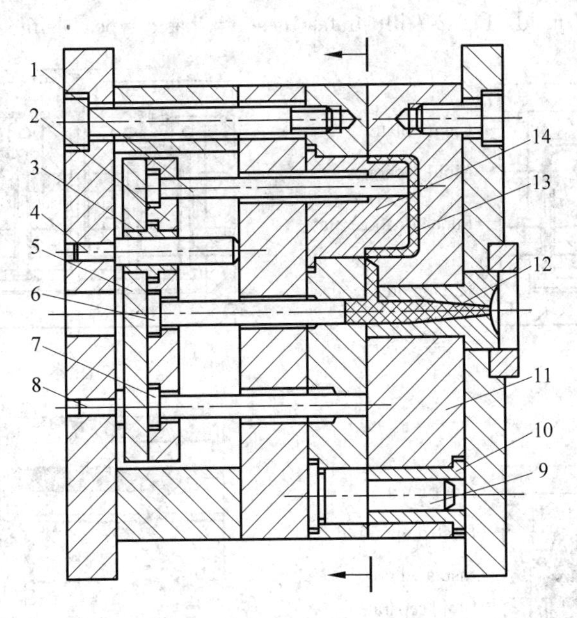

Fig. 2-6 Injection mold

1—ejector pin; 2—ejector plate; 3—guide bush; 4—guide pillar; 5—ejector base plate; 6—sprue puller pin; 7—return pin; 8—limit pin; 9—guide pillar; 10—guide bush; 11—cavity plate; 12—sprue bushing; 13—plastic workpiece; 14—core

separating line between the two mold halves is called the parting line. The injected material is transferred through a central feed channel, called the sprue. The sprue is located on the sprue bushing and is tapered to facilitate release of the sprue material from the mold during mold opening. In multi-cavity molds, the sprue feeds the polymer melt to a runner system, which leads into each mold cavity through a gate.

The core plate holds the main core. The purpose of the main core is to establish the inside configuration of the part. The core plate has a backup or support plate. The support plate in turn is supported by pillars against the U-shaped structure known as the ejector housing, which consists of the rear clamping plate and spacer blocks. This U-shaped structure, which is bolted to the core plate, provides the space for the ejection stroke also known as the stripper stroke. During solidification the part shrinks around the main core so that when the mold opens, part and sprue are carried along with the moving mold half. Subsequently, the central ejector is activated, causing the ejector plates to move forward so that the ejector pins can push the part off the core. Both mold halves are provided with cooling channels through which cooled water is circulated to absorb the heat delivered to the mold by the hot thermoplastic polymer melt. The mold cavities also incorporate fine vents (0.02 to 0.08 mm by 5 mm) to ensure that no air is trapped during filling.

There are six basic types of injection molds in use today. They are: (a) two-plate mold; (b) three-plate mold, (c) hot-runner mold; (d) insulated hot-runner mold; (e) hot-manifold mold; and (f) stacked mold. Fig. 2-7 illustrates these six basic types of injection molds.

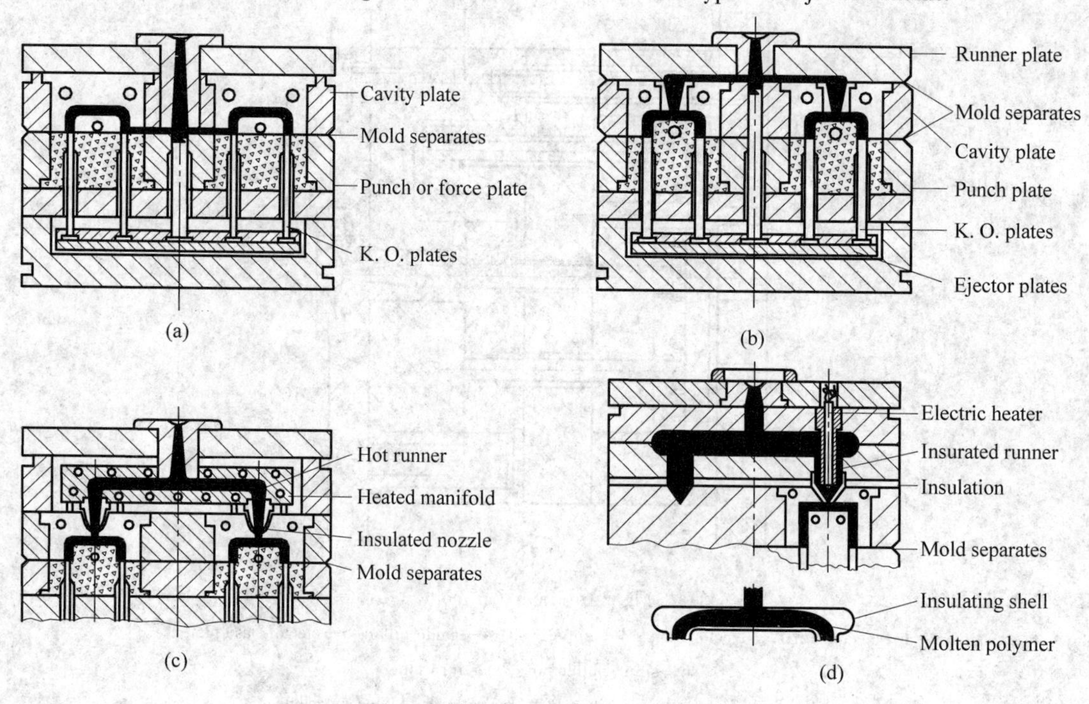

Fig. 2-7 Six basic types of injection molding dies

(a) Two-plate mold; (b) Three-plate mold; (c) Hot-runner mold; (d) Insulated runner mold; (e) Hot manifold mold; (f) Stacked mold

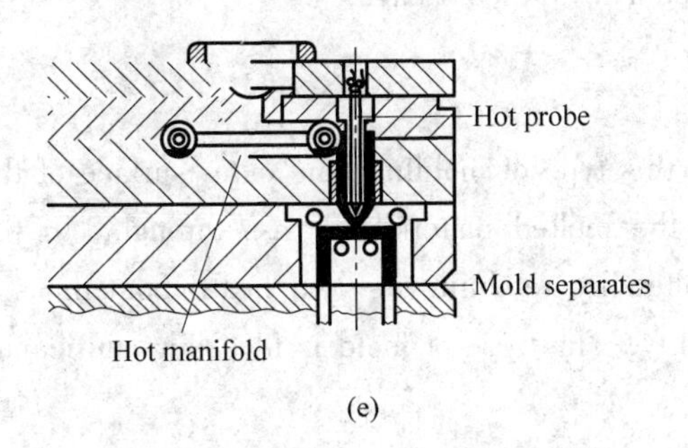

(e)

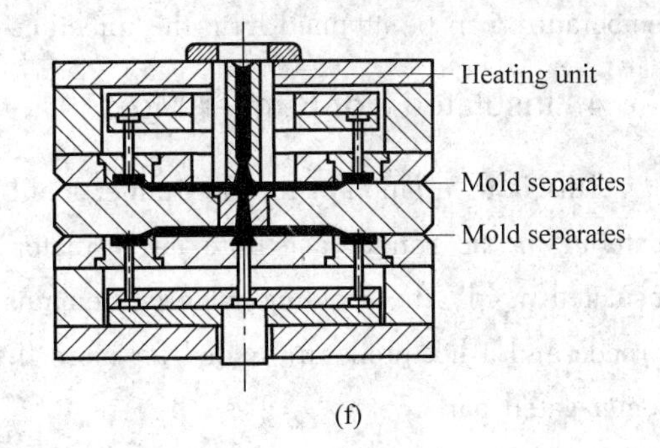

(f)

Fig. 2-7 (Continued)

1. Two-Plate Mold

A two-plate mold consists of two plates with the cavity and cores mounted in either plate. The plates are fastened to the press platens. The moving half of the mold usually contains the ejector mechanism and the runner system. All basic designs for injection molds have this design concept. A two-plate mold is the most logical type of tool to use for parts that require large gates.

2. Three-Plate Mold

This type of mold is made up of three plates: (1) the stationary or runner plate is attached to the stationary platen, and usually contains the sprue and half of the runner; (2) the middle plate or cavity plate, which contains half of the runner and gate, is allowed to float when the mold is open; and (3) the movable plate or force plate contains the molded part and the ejector system for the removal of the molded part. When the press starts to open, the middle plate and the movable plate move together, thus releasing the sprue and runner system and degating the molded part. This type of mold design makes it possible to segregate the runner system and the part when the mold opens. The die design makes it possible to use center-pin-point gating.

3. Hot-Runner Mold

In this process of injection molding, the runners are kept hot in order to keep the molten plastic in a fluid state at all times. In effect this is a "runnerless" molding process and is sometimes called the same. In runnerless molds, the runner is contained in a plate of its own. Hot runner molds are similar to three-plate injection molds, except that the runner section of the mold is not opened during the molding cycle. The heated runner plate is insulated from the rest of the cooled mold. Other than the heated plate for the runner, the remainder of the mold is a standard two-plate die.

Runnerless molding has several advantages over conventional sprue runner-type molding. There are no molded side products (gates, runners, or sprues) to be disposed of or reused, and there is no separating of the gate from the part. The cycle time is only as long as is required for the molded part to be cooled and ejected from the mold. In this system, a uniform melt temperature can be attained from the injection cylinder to the mold cavities.

4. Insulated Hot-Runner Mold

This is a variation of the hot-runner mold. In this type of molding, the outer surface of the material in the runner acts like an insulator for the melted material to pass through. In the insulated mold, the molding material remains molten by retaining its own heat. Sometimes a torpedo and a hot probe are added for more flexibility. This type of mold is ideal for multicavity center-gated parts.

5. Hot-Manifold

This is a variation of the hot-runner mold. In the hot-manifold die, the runner but not the runner plate is heated. This is done by using an electric-cartridge-insert probe.

6. Stacked Mold

The stacked injection mold is just what the name implies. A multiple two-plate mold is placed one on top of the other. This construction can also be used with three-plate molds and hot-runner molds. A stacked two-mold construction doubles the output from a single press and reduces the clamping pressure required to one half, as compared to a mold of the same number of cavities in a two-plate mold. This method is sometimes called "two-level molding".

2.3.3 Mold Machine

1. Conventional Injection-Molding Machine

In this process, the plastic granules or pellets are poured into a machine hopper and fed into the chamber of the heating cylinder. A plunger then compresses the material, forcing it through progressively hotter zones of the heating cylinder, where it is spread thin by a torpedo. The torpedo is installed in the center of the cylinder in order to accelerate the heating of the center of the plastic mass. The torpedo may also be heated so that the plastic is heated from the inside as well as from the outside.

The material flows from the heating cylinder through a nozzle into the mold. The nozzle is the seal between the cylinder and the mold; it is used to prevent leaking of material caused by the pressure used. The mold is held shut by the clamp end of the machine. For polystyrene, two to three tons of pressure on the clamp end of the machine is generally used for each inch of projected area of the part and runner system. The conventional plunger machine is the only type of machine that can produce a mottle-colored part. The other types of injection machines mix the plastic material so thoroughly that only one color will be produced.

2. Piston-Type Preplastifying Machine

This machine employs a torpedo ram heater to preplastify the plastic granules. After the melt stage, the fluid plastic is pushed into a holding chamber until it is ready to be forced into the die. This type of machine produces pieces faster than a conventional machine, because the molding chamber is filled to shot capacity during the cooling time of the part. Due to the fact that the injection plunger is acting on fluid material, no pressure loss is encountered in compacting the granules. This allows for larger parts with more projected area. The remaining features of a piston-type preplastifying machine are identical to the conventional single-plunger injection machine. Fig. 2-8 illustrates a piston or plunger preplastifying injection molding machine.

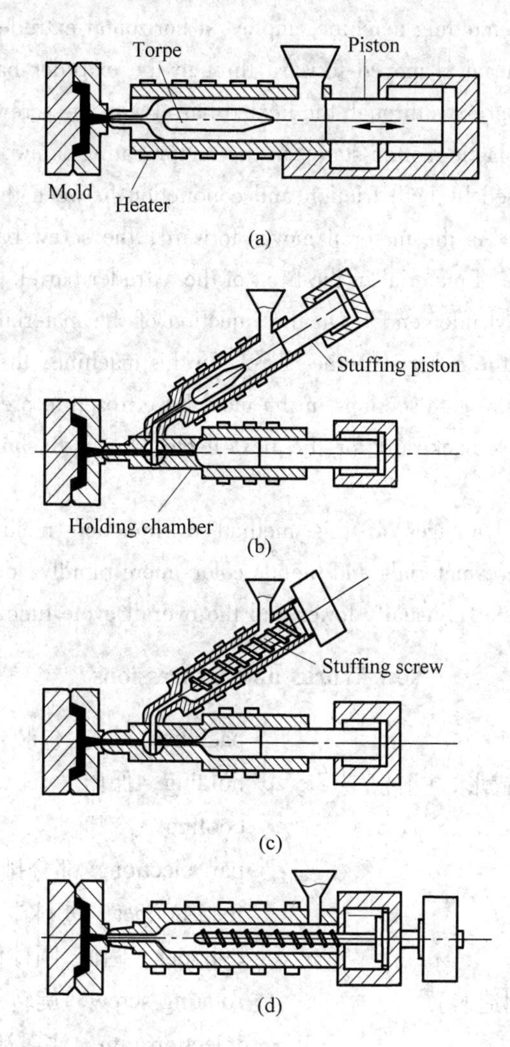

Fig. 2-8 The four basic types of injection molding equipment

(a) conventional injection machine; (b) preplastifying machine piston type; (c) reciprocating screw machine; (d) reciprocating screw machine

3. Screw-Type Preplastifying Machine

In this injection-molding machine, an extruder is used to plasticize the plastic material. The turning screw feeds the pellets forward to the heated interior surface of the extruder barrel. The molten, plasticized material moves from the extruder into a holding chamber, and from there is forced into the die by the injection plunger. The use of a screw gives the following advantages: (1) better mixing and shear action of the plastic melt; (2) a broader range of stiffer flow and heat-sensitive materials can be run; (3) color changes can be handled in a shorter time, and (4) fewer stresses are obtained in the molded part.

4. Reciprocating-Screw Injection Machine

This type of injection molding machine employs a horizontal extruder in place of the heating chamber. The plastic material is moved forward through the extruder barrel by the rotation of a screw. As the material progresses through the heated barrel with the screw, it is changing from the granular condition to the plastic molten state. In the reciprocating screw, the heat delivered to the molding compound is caused by both friction and conduction between the screw and the walls of the barrel of the extruder. As the material moves forward, the screw backs up to a limit switch that determines the volume of material in the front of the extruder barrel. It is at this point that the resemblance to a typical extruder ends. On the injection of the material into the die, the screw moves forward to displace the material in the barrel. In this machine, the screw performs as a ram as well as a screw. After the gate sections in the mold have frozen to prevent backflow, the screw begins to rotate and moves backward for the next cycle. Fig. 2-8 shows a reciprocating-screw injection machine.

There are several advantages to this method of injection molding. It more efficiently plasticizes the heat-sensitive materials and blends colors more rapidly, due to the mixing action of the screw. The material heat is usually lower and the overall cycle time is shorter.

New Words and Expressions

injection mold 注射模
injection molding 注射成形,注射模塑
container 容器
housing 机架
tool handle 工具柄
knob 旋钮,球形捏手
plumbing fitting 铅管制造装置
mold wall 模壁
grain size 晶粒度
grain orientation 晶粒取向
foamed part 泡沫部分
reaction injection 反应注射
injection filling 注射填充
packing 填充,包装
holding 保压
cooling 冷却
part ejection 部分排除物
feed hopper 进料斗
cylinder 圆筒,圆柱体
rotating screw 旋转螺杆
injection ram 注射活塞
barrel 料筒
nozzle 管口,喷嘴
sprue 浇道,主流道
runner system 流道系统
mold cavity 模具型腔

volumetrically 容积地
skin layer 表层
core 芯子,型芯
piston 活塞
injection chamber 注射室
torpedo 分料梭
vent 出口,通风孔
ejector pin 起模杆,顶杆
reciprocate 互换
defect 过失,缺陷
sink mark 收缩痕
void 气孔,中空的
dimensional instability 尺寸不稳定性
encompass 包围,环绕
packing pressure 保压压力
ejecting temperature 脱模温度,取出温度
plasticate 塑炼,塑化(橡胶等)(指通过加热及挤压使其粒子软化)
shearing action 剪切作用
heater band 电热丝
cavity plate 型腔固定板
core plate 中心板
parting line 分型线
main core 主型芯
backup 阻塞,后援
support plate 支撑板
U-shaped structure U形结构
ejector housing 注塑模顶杆空间
spacer block 垫块
two-plate mold 双板模
press platen 压板
moving half of mold 瓣合机构
ejector mechanism 顶出机构
three-plate mold 三板模
stationary platen 固定压盘
middle plate 中间板
segregate 隔离
center-pin-point gating 点浇口浇注系统
hot-runner mold 热流道模具
runnerless molding 无流道模具
molding cycle 成形周期
remainder 残余,剩余物
side product 副产品,废料
cycle time 周期
melt temperature 熔化温度
insulated hot-runner 绝热保温流道
insulator 绝缘体,绝缘器
hot-manifold mold 热流道模具
hot probe 热探测器
multicavity 多腔
multicavity center-gated 多腔中心浇口的
runner plate 流道板
stacked mold 重叠压塑模具
clamping 合模
two-level molding 双层模塑
granule 颗粒,细粒
mottle-colored part 斑点部分
priston-typed preplastifying machine 柱塞式预塑机
reciprocating-screw injecting machine 往复式螺杆注塑机
holding chamber 储料室
extruder 挤压机
limit switch 限位开关,行程开关
backflow 逆流
blend 混合

2.4 Other Molding Processes

2.4.1 Compression Molding

Compression molding is the basic forming process where an appropriate amount of material is introduced into a heated mold, which is subsequently closed under pressure. The molding material, softened by heat, is formed into a continuous mass having the geometrical configuration of the mold cavity. Further heating (thermosetting plastics) results in hardening of the molding material. If thermoplastics are the molding material, hardening is accomplished by cooling the mold.

Fig. 2-9 illustrates types of compression molding. Here the molding compound is placed in the heated mold. After the plastic compound softens and becomes plastic, the punch moves down and compresses the material to the required density by a pressure. Some excess material will flow (vertical flash) from the mold as the mold closes to its final position.

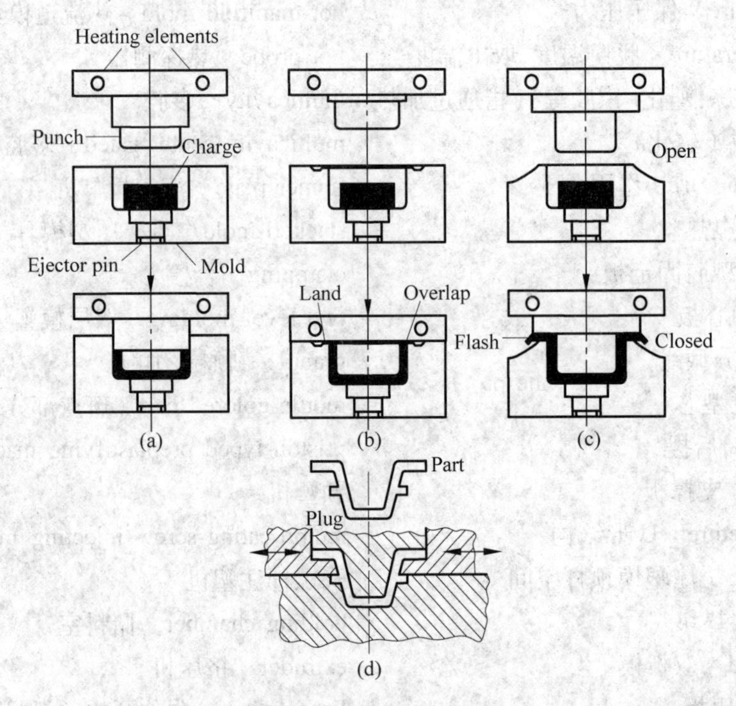

Fig. 2-9 Types of compression molding

(a) positive; (b) semipositive; (c) flash; (d) die design for making a compression-molded part with undercuts

Continued heat and pressure produce the chemical reaction which hardens the compound. The time required for polymerization or curing depends principally upon the largest cross section of the product and the type of molding compound. The time may be less than a minute, or it may take several minutes before the part is ejected from the cavity.

Since the plastic material is placed directly into the mold cavity, the mold itself can be simpler than those used for other molding processes. Gates and sprues are unnecessary. This also results in a saving in material, because trimmed-off gates and sprues would be a complete loss of the thermosetting plastic. The press used for compression molding is usually a vertical hydraulic press. Large presses may require the full attention of one operator. However, several smaller presses can be operated by one operator. The presses are conveniently located so the operator can easily move from one to the next. By the time he gets around to a particular press again, that mold will be ready to open.

The thermosetting plastics which harden under heat and pressure are suitable for compression molding and transfer molding. It is not practical to mold thermoplastic materials by these methods, since the molds would have to be alternately heated and cooled. In order to harden and eject thermoplastic parts from the mold, cooling would be necessary.

2.4.2 Transfer Molding

The transfer molding process consists of placing a charge of material (extrudate or preheated preform) into the chamber, referred to as the pot. The press is activated and travels upward making contact with the floating plate, which closes the two halves of the mold.

Further travel of both plates causes contact of the plunger with the material in the pot. Material is then forced through a sprue or sprues directly into the closed cavity. When the cavity is completely filled, the excess material forms a cull in the pot (excess waste material). After the part is cured, the press is opened and the floating plate and bottom plate separate from the top plate, exposing the plunger and cull. As the press travel continues, the floating plate motion is stopped by straps fastened to the top plate.

This separates the two halves of the mold, and the part remains in the lower half until knockout pins extract it. Since the process requires that the single charge (shot) of material be transferred from the pot to the cavities, it is known as pot-type transfer (Fig. 2-10). An operator is needed to remove the cull from the pot plunger, remove the part or parts, clean the mold, charge a single shot of preheated material into the pot area, and activate the press.

A relatively short time after the patenting of the transfer mold; transfer presses were developed. These consist of a main clamping ram located at either the top or the bottom of the press, with one or more auxiliary rams mounted opposite the clamping ram. The clamping rams activate the movable platen. The auxiliary rams are fastened to the stationary platen and are used to activate a plunger, which moves within a transfer sleeve or cylinder. For the plunger in the bottom half of the mold, the process consists of placing preheated preforms or extrudates in the transfer sleeve or cylinder, closing the two halves of the mold, and activating the plunger, which forces material out through channels, known as runners, and through the restricted gate area into the mold halves. When the cavities are completely filled, the excess material remains as a cull at the face of the plunger. After the material is cured, the press is opened at the parting line, parts are removed and the gate, runner and cull. This molding process is commonly called the plun ger-transfer method. A typical mold construction is shown in Fig. 2-11. If the bottom plunger- transfer mold is constructed, the operation may be automated, since auxiliary devices may load the preheated preforms, and unloading trays may be utilized to receive and separate the parts, runner, gates, and plunger culls. In all other cases an operator is required for each press.

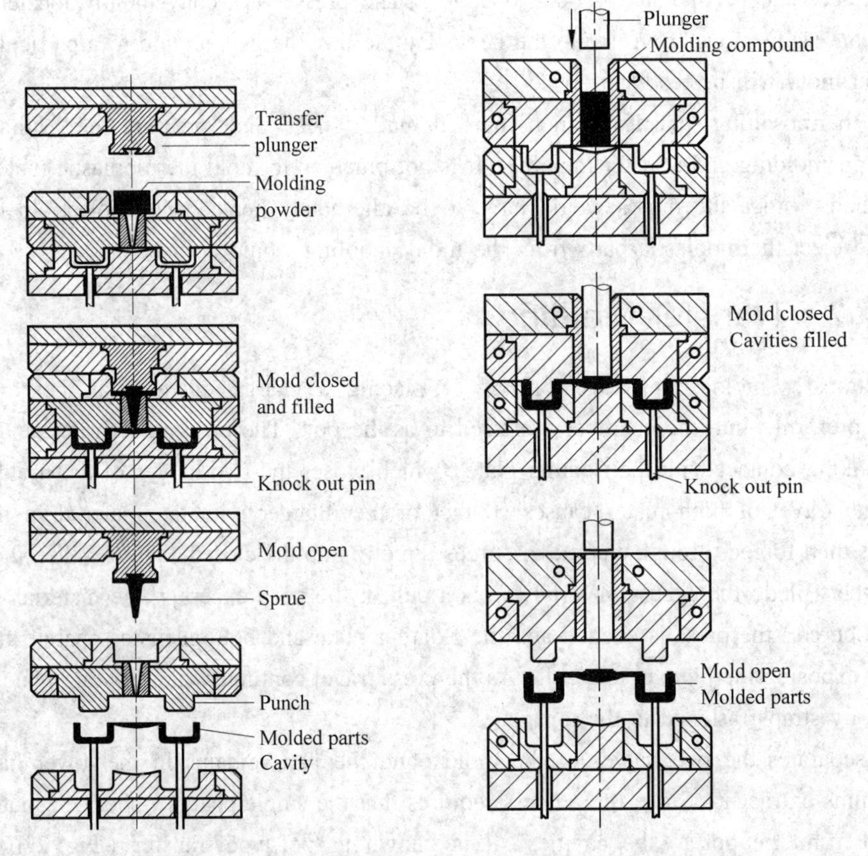

Fig. 2-10 The pot-type pot or sprue type transfer molding

Fig. 2-11 The plunger-transfer molding

The two-stage plunger transfer process requires a conventionally designed hydraulic or toggle top clamp press, with a bottom transfer cylinder and plunger. A reciprocal screw within a heated barrel is mounted horizontally next to the press. The granular material charge is preheated in the barrel and is discharged into the transfer cylinder or sleeve through an opening in its side. The material flow from the same way as described in the plunger-transfer molding process.

The two-stage plunger-transfer molds are similar in construction to the plunger transfer, except that a special transfer cylinder or sleeve and plunger are required.

2.4.3 Co-injection Molding

Co-injection molding is used to produce parts that have a laminated structure with the core material embedded between the layers of the skin material. As shown in Fig. 2-12, the process involves sequential injection of two different bur compatible polymer melts into a cavity where the materials laminate and solidify. A shot of skin polymer melt is first injected into the mold (Fig. 2-12(a)), followed by core polymer melt which is injected until the mold cavity is nearly filled (Fig. 2-12(b)); the skin polymer is then injected again to purge the core polymer away from the sprue (Fig. 2-12(c)). The process offers the inherent flexibility of using the optimal properties of each material or modifying the properties of each material or those of the molded part.

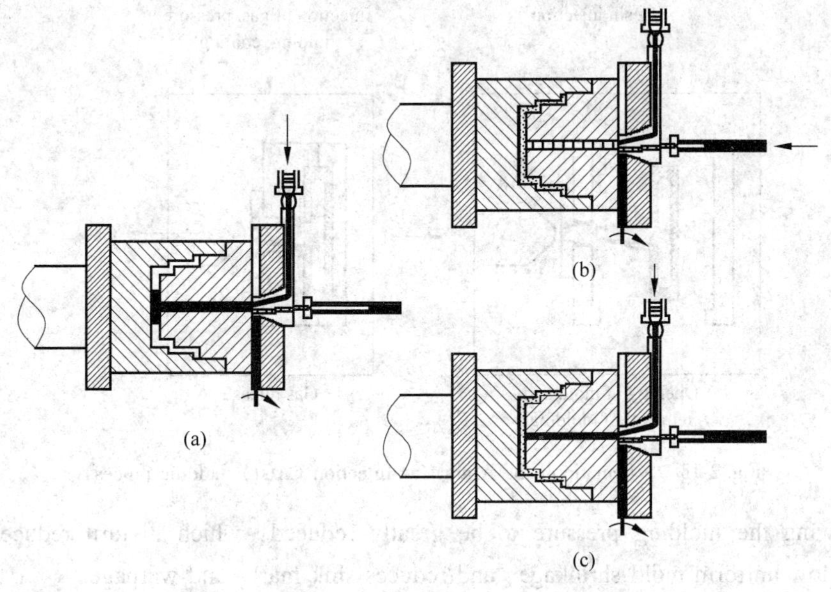

Fig. 2-12 Three stages of co – injection (sandwich) molding

(a) short shot of skin polymer melt (shown in black) is injected into the mold;

(b) injection of core polymer melt until cavity is nearly filled;

(c) skin polymer melt is injected again, pushing the core polymer away from the sprue

2.4.4 Gas-Assisted Injection Molding

The gas-assisted injection molding process begins with a partial for full injection of polymer melt into the mold cavity. Compressed gas is then injected into the core of the polymer melt to help fill and pack the mold, as shown in Fig. 2-13 for the Asahi Gas Injection Molding process. This process is thus capable of producing hollow rigid parts, free of sink marks. The hollowing out of thick sections of moldings results in reduction part weight and saving of resin material.

Other advantages included shorter cooling cycles, reduced clamp force tonnage and part consolidation. The process allows high precision molding with greater dimensional stability by eliminating uneven mold shrinkage and makes it possible to mold complicated shapes in single form, thus reducing product assembly work and simplifying mold design.

The formation of thick walled sections of a molding can be easily achieved by introducing gas in the desired locations. The gas channels thus formed also effectively support the flow of

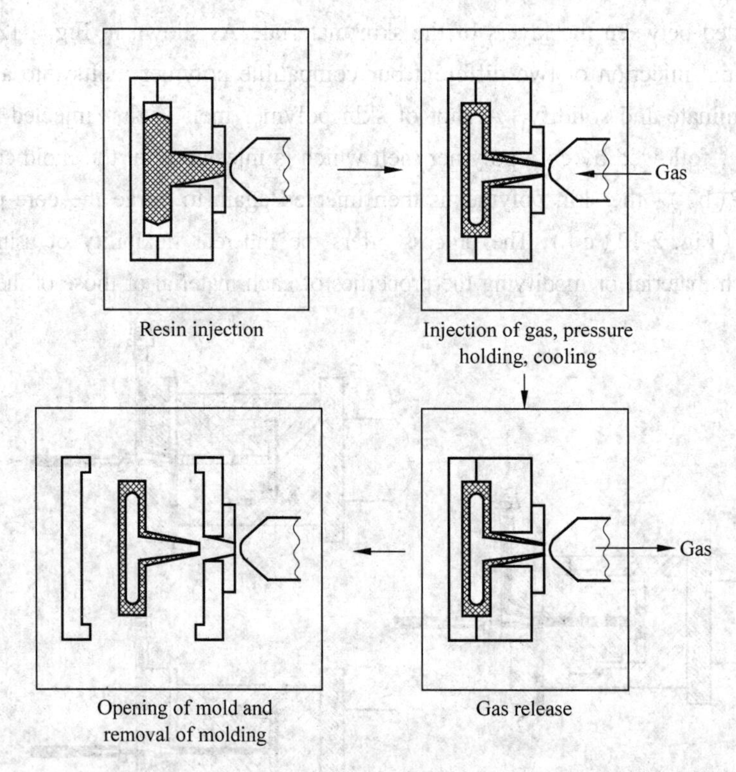

Fig. 2-13 Schematic of the Asahi Gas Injection (AGI) molding process

resin, allowing the molding pressure to be greatly reduced, which in turn reduces internal stresses, allow uniform mold shrinkage, and reduces sink marks and warpage.

2.4.5 Extrusion Molding

The extrusion process is basically designed to continuously convert a soft material into a particular form. An oversimplified analogy may be a house-hold meat grinder. However, unlike the extrudate from a meat grinder, plastic extrudates generally approach truly continuous formation. Like the usual meat grinder, the extruder (Fig. 2-14) is essentially a screw conveyor. It carries the cold plastic material (in granular or powered form) forward by the action of the screw, squeezes it, and, with heat from external heaters and the friction of viscous flow, changes it to a molten stream. As it does this, it develops pressure on the material, which is highest right before the molten plastic enters the die. The screen pack, consisting of a number of fine or coarse mesh gauzes supported on a breaker plate and placed between the screw and the die, filter out dirt and unfused polymer lumps. The pressure on the molten plastic forces it through an adapter and into the die, which dictates the shape of the fine extrudate. A die with a round opening as shown in Fig. 2-14, produces pipe; a square die opening produces a square profile, etc. Other continuous shapes, such as the film, sheet, rods, tubing, and filaments, can be produced with appropriate dies. Extruders are also used to apply insulation and jacketing to wire and cable and to coat substrates such as paper, cloth, and foil.

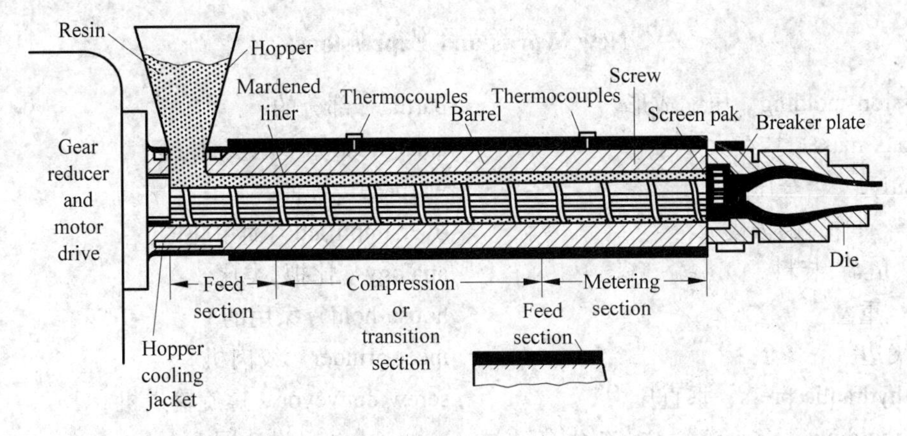

Fig. 2-14 Scheme for a typical single - screw extruder showing extruding pipe

When thermoplastic polymers are extruded, it is necessary to cool the extrudate below T_m or T_g to impart dimensional stability. This cooling can often be done simply by running the product through a tank of water, by spraying cold water, or, even more simply, by air cooling. When rubber is extruded, dimensional stability results from cross-linking (vulcanization). Interestingly, extrusion for wire coating was the first application of the screw extruder in polymer processing.

2.4.6 Blow Molding

Basically, blow molding is intended for use in manufacturing hollow plastic products, such as bottles and other containers. However, the process is also used for the production of toys, automobile parts, accessories, and many engineering components. The principles used in blow molding are essentially similar to those used in the production of glass bottles. Although there are considerable differences in the process available for blow molding, the basic steps are the same: (1) melt the plastic; (2) form the molten plastic into a parison (a tubelike shape of molten plastic); (3) seal the ends of the parison except for one area through which the blowing air can enter; (4) inflate the parison to assume the shape of the mold in which it is placed; (5) cool the blow-molded part; (6) eject the blow-molded part; (7) trim flash if necessary.

Two basic processes of blow molding are extrusion blow molding and injection blow molding. These processes differ in the way in which the parison is made. The extrusion process utilizes an unsupported parison, whereas the injection process utilizes a parison supported on a metal core. The extrusion blow-molding process by far accounts for the largest percentage of blow-molded objects produced today. The injection process is, however, gaining acceptance.

Although any thermoplastic can be blow-molded, polyethylene products made by this technique are predominant. Polyethylene squeeze bottles form a large percentage of all blow-molded products.

New Words and Expressions

compression molding 压缩成形
continuous mass 持续大量
semipositive 半全压式
undercut 凹槽,倒拔模
charge 负荷,装料
overlap 重叠
flash 飞边
vertical hydraulic press 垂直压力
transfer molding 传递模塑法,连续自动送进成形
extrudate 挤出物
preheated perform 预热型坯
floating plate (平板机的)中间热板
cull 精选
strap 带,皮带,用带捆扎
knockout pin 顶出杆
auxiliary ram 辅助活塞
transfer sleeve 传递套筒
plunger-transfer 传递活塞
toggle 套索钉,拴牢
co-injection molding 共注射成形
laminated 分层的
embedded 嵌入的,内嵌的
purge 清除,清洗
gas-assisted 气体辅助的
partial 部分的
consolidation 压实,压密
warpage 翘曲
oversimplified 简单化的
analogy 比拟,类比
house-hold 家用的
meat grinder 绞肉机
screw conveyor 螺旋输送机
screen pack 过滤网
mesh gauze 筛网过滤器
breaker plate 机头体
thermocouple 热电偶
hopper 料斗
gear reducer 齿轮减速装置
motor drive 电机驱动
lump 团,块
dictate 控制,支配
tubing 管道
filament 细丝
foil 铝箔
tank 储水池
automobile parts 汽车零部件
accessories 配件
inflate 使充气,膨胀

Casting Dies

3.1 Casting

The first castings were made during the period 4000 ~ 3000 B. C. , using stone and metal molds for casting copper. Various casting processes have been developed over a long period of time, each with its own characteristics and applications, to meet specific engineering and service requirements. Many parts and components are made by casting, including cameras, carburetors, engine blocks, crankshafts, automotive components, agricultural and railroad equipment, pipes and plumbing fixtures, power tools, gun barrels, frying pans, and very large components for hydraulic turbines.

Casting can be done in several ways. The two major ones are sand casting, in which the molds used are disposable after each cycle, and die casting, or permanent molding, in which the same metallic die is used thousands or even millions of times. Both types of molds have three common features. They both have a "plumbing" system to channel molten alloy into the mold cavity. These channels are called sprues, runners, and gates (Fig. 3-1). Molds may be modified by cores which form holes and undercuts or inserts that become an integral part of the casting. Inserts strengthen and reduce friction, and they may be more machinable than the surrounding metal. For example, a steel shaft when properly inserted into a die cavity results in an assembled aluminum step gear after the shot.

After pouring or injection, the resulting castings require subsequent operations such as trimming, inspection, grinding, and repairs to a greater or lesser extent prior to shipping. Premium-quality castings from alloys of aluminum or steel require x-ray soundness that will be acceptable by the customer.

Certain special casting processes are precision-investment casting, low-pressure casting, and centrifugal casting.

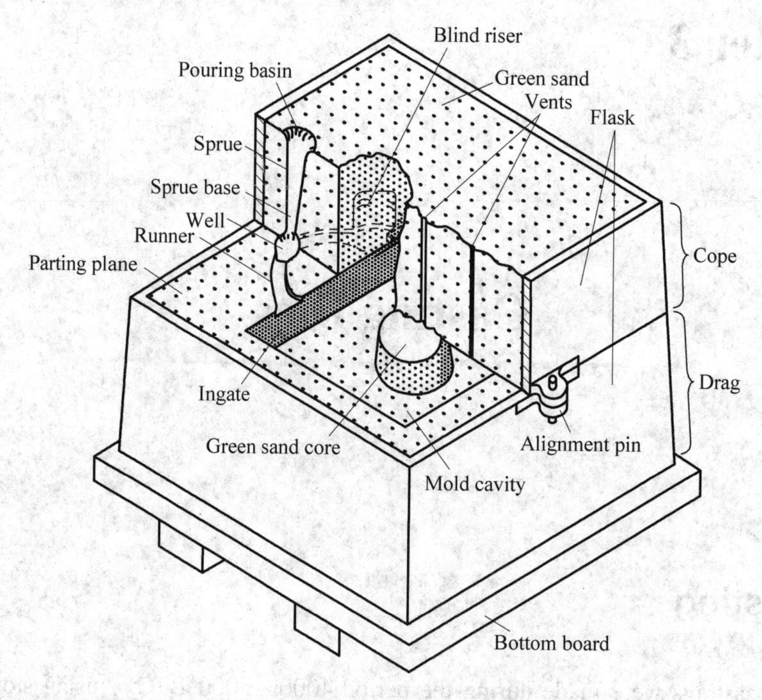

Fig. 3-1 A typical mold partially sectioned to show detail

New Words and Expressions

metal mold 金属模
copper 铜
camera 照相机
carburetor 汽化器
engine block 发动机组(本体)
crankshaft 曲轴,曲柄轴
automotive component 自动推进部件
plumbing fixture 管道夹具
power tool 电动工具
gun barrel 炮筒,枪筒
frying pan 煎锅,长柄平锅
sand casting 砂型铸造
channel 流道
runners 分流道
gate 浇口
machinable 可加工的
steel shaft 钢杆
aluminum 铝
pouring 浇注
shipping 发货,装运
premium-quality 第一流的质量
X-ray X 射线
precision-investment casting 精密熔模铸造
low-pressure casting 低压铸造
centrifugal casting 离心铸造法

3.2 Sand Casting

The traditional method of casting metals is in sand molds and has been used for millennia. Simply stated, sand casting consists of (a) placing a pattern having the shape of the desired casting in sand to make an imprint, (b) incorporating a gating system, (c) filling the resulting cavity with molten metal, (d) allowing the metal to cool until it solidifies, (e) breaking away the sand mold, and (f) removing the casting (Fig. 3-2). The production steps for a typical sand-casting operation are shown in Fig. 3-3.

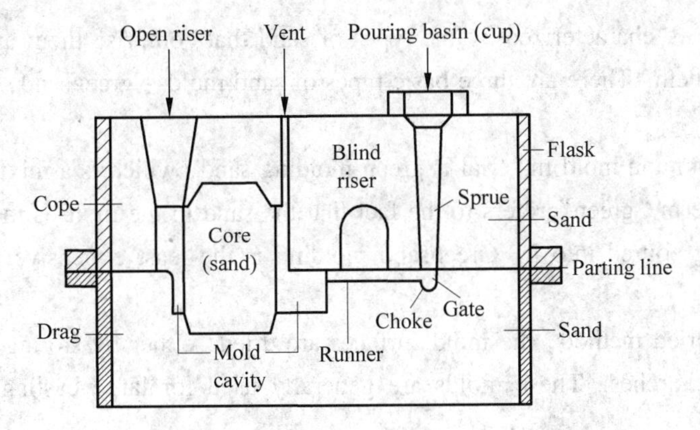

Fig. 3-2 Schematic illustration of a sand mold

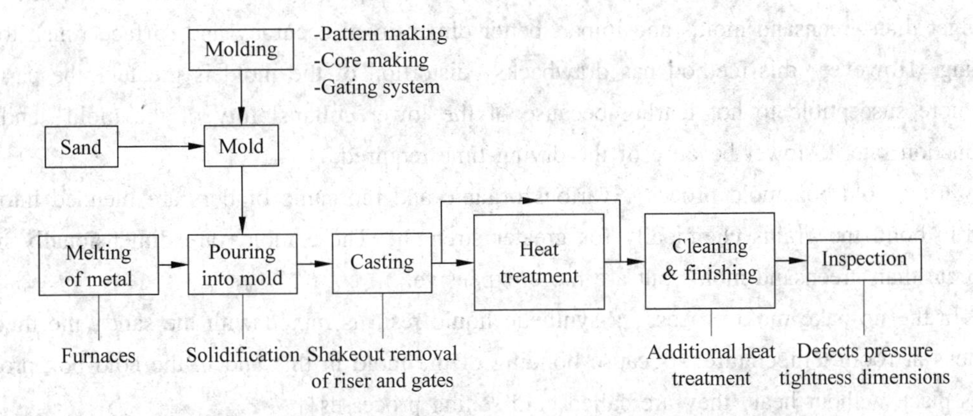

Fig. 3-3 Outline of production steps in a typical sand-casting operation

Although the origins of sand casting date to ancient times, it is still the most prevalent form of casting. In the United States alone, about 15 million tons of metal are cast by this method each year.

3.2.1 Sands

Most sand casting operations use silica sand (SiO_2), which is the product of the disintegration of rocks over extremely long periods of time. Sand is inexpensive and is suitable as mold material because of its resistance to high temperatures. There are two general types of sand: naturally bonded (bank sand) and synthetic (lake sand). Because its composition can be controlled more accurately, synthetic sand is preferred by most foundries.

Several factors are important in the selection of sand for molds. Sand having fine, round grains can be closely packed and forms a smooth mold surface. Although fine-grained sand enhances mold strength, the fine grains also lower mold permeability. Good permeability of molds and cores allows gases and steam evolved during casting to escape easily.

3.2.2 Types of Sand Molds

Sand molds are characterized by the types of sand that comprise them and by the methods used to produce them. There are three basic types of sand molds: greensand, cold-box, and no-bake molds.

The most common mold material is green molding sand, which is a mixture of sand, clay, and water. The term "green" refers to the fact that the sand in the mold is moist or damp while the metal is being poured into it. Greensand molding is the least expensive method of making molds.

In the skin-dried method, the mold surfaces are dried, either by storing the mold in air or by drying it with torches. These molds are generally used for large castings because of their higher strength.

Sand molds are also oven dried (baked) prior to pouring the molten metal; they are stronger than greensand molds and impart better dimensional accuracy and surface finish to the casting. However, this method has drawbacks: distortion of the mold is greater; the castings are more susceptible to hot tearing because of the lower collapsibility of the mold; and the production rate is slower because of the drying time required.

In the cold-box mold process, various organic and inorganic binders are blended into the sand to bond the grains chemically for greater strength. These molds are dimensionally more accurate than greensand molds but are more expensive.

In the no-bake mold process, a synthetic liquid resin is mixed with the sand; the mixture hardens at room temperature. Because bonding of the mold in this and in the cold-box process takes place without heat, they are called cold-setting processes.

The following are the major components of sand molds (Fig. 3-2):

(1) The mold itself, which is supported by a flask. Two-piece molds consist of a cope on top and a drag on the bottom. The seam between them is the parting line. When more than two pieces are used, the additional parts are called cheeks.

(2) A pouring basin or pouring cup, into which the molten metal is poured.

(3) A sprue, through which the molten metal flows downward.

(4) The runner system, which has channels that carry the molten metal from the sprue to the mold cavity. Gates are the inlets into the mold cavity.

(5) Risers, which supply additional metal to the casting as it shrinks during solidification. Fig. 3-2 shows two different types of risers: a blind riser and an open riser.

(6) Cores, which are inserts made from sand. They are placed in the mold to form hollow regions or otherwise define the interior surface of the casting. Cores are also used on the outside of the casting to form features such as lettering on the surface of a casting or deep external pockets.

(7) Vents, which are placed in molds to carry off gases produced when the molten metal

comes into contact with the sand in the mold and core. They also exhaust air from the mold cavity as the molten metal flows into the mold.

3.2.3 Patterns

Patterns are used to mold the sand mixture into the shape of the casting. They may be made of wood, plastic, or metal. The selection of a pattern material depends on the size and shape of the casting, the dimensional accuracy, the quantity of castings required, and the molding process.

Because patterns are used repeatedly to make molds, the strength and durability of the material selected for patterns must reflect the number of castings that the mold will produce. They may be made of a combination of materials to reduce wear in critical regions. Patterns are usually coated with a parting agent to facilitate their removal from the molds.

Patterns can be designed with a variety of features to fit application and economic requirements. One-piece patterns, also called loose or solid patterns, are generally used for simpler shapes and low-quantity production. They are generally made of wood and are inexpensive. Split patterns are two-piece patterns made such that each part forms a portion of the cavity for the casting; in this way, castings with complicated shapes can be produced.

Match-plate patterns are a popular type of mounted pattern in which two-piece patterns are constructed by securing each half of one or more split patterns to the opposite sides of a single plate (Fig. 3-4). In such constructions, the gating system can be mounted on the drag side of the pattern. This type of pattern is used most often in conjunction with molding machines and large production runs to produce smaller castings.

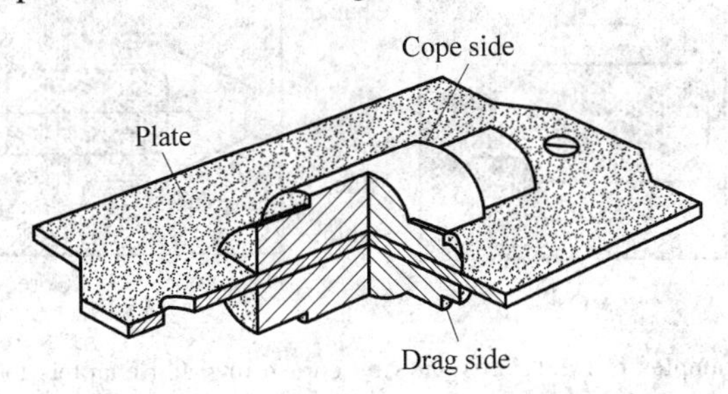

Fig. 3-4 A typical metal match-plate pattern used in sand casting

An important recent development is the application of rapid prototyping to mold and pattern making. In sand casting, for example, a pattern can be fabricated in a rapid prototyping machine and fastened to a backing plate at a fraction of the time and cost of machining a pattern. There are several rapid prototyping techniques with which these tools can be produced quickly.

Pattern design is a crucial aspect of the total casting operation. The design should provide for metal shrinkage, case of removal from the sand mold by means of a taper or draft (Fig. 3-5), and proper metal flow in the mold cavity.

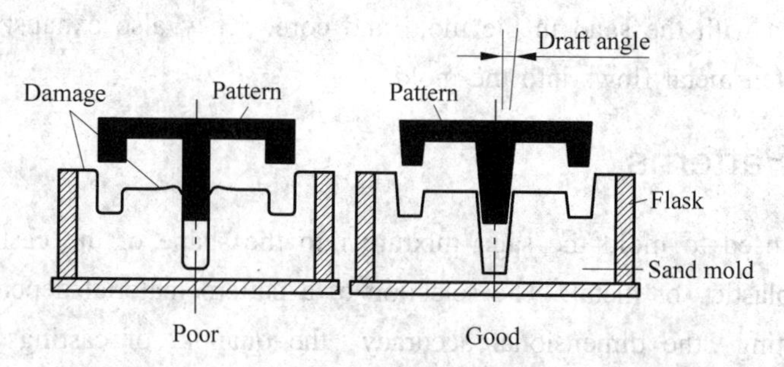

Fig. 3-5 Taper on patterns for case of removal from the sand mold

3.2.4 Cores

For castings with internal cavities or passages, such as those found in an automotive engine block or a valve body, cores are utilized. Cores are placed in the mold cavity before casting to form the interior surfaces of the casting and are removed from the finished part during shakeout and further processing. Like molds, cores must possess strength, permeability, ability to withstand heat, and collapsibility; therefore, cores are made of sand aggregates.

The core is anchored by core prints. These are recesses that are added to the pattern to support the core and to provide vents for the escape of gases (Fig. 3-6). A common problem with cores is that for some casting requirements, as in the case where a recess is required, they may lack sufficient structural support in the cavity. To keep the core from shifting, metal supports (chaplets) may be used to anchor the core in place (Fig. 3-6).

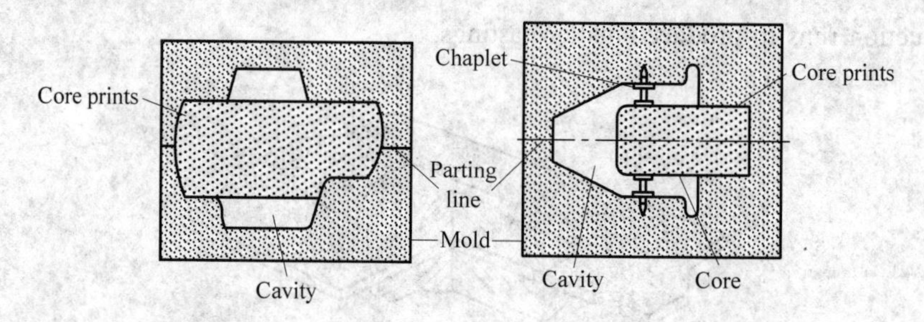

Fig. 3-6 Examples of sand cores showing core prints and chaplets to support cores

Cores are generally made in a manner similar to that used in making molds; the majority are made with shell, no-bake, or cold-box processes. Cores are formed in core boxes, which are used in much the same way that patterns are used to form sand molds. The sand can be packed into the boxes with sweeps, or blown into the box by compressed air from core blowers. The latter have the advantages of producing uniform cores and operating at very high production rates.

3.2.5 Sand-Molding Machines

The oldest known method of molding, which is still used for simple castings, is to compact the sand by hand hammering (tamping) or ramming it around the pattern. For most operations, however, the sand mixture is compacted around the pattern by molding machines (Fig. 3-7). These machines eliminate arduous labor, offer high-quality casting by improving the application and distribution of forces, manipulate the mold in a carefully controlled manner, and increase production rate.

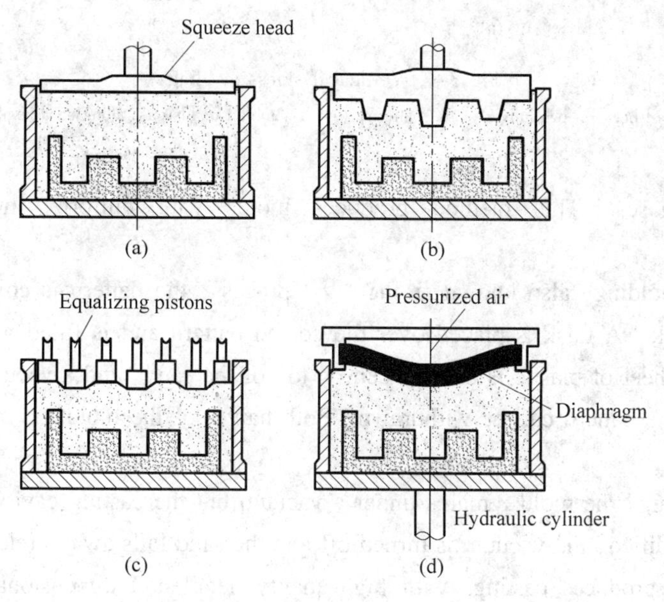

Fig. 3-7 Various designs of squeeze heads for mold making
(a) conventional flat head; (b) profile head; (c) equalizing squeeze pistons; (d) flexible diaphragm

Mechanization of the molding process can be further assisted by jolting the assembly. The flask, molding sand, and pattern are first placed on a pattern plate mounted on an anvil, and then jolted upward by air pressure at rapid intervals. The inertial forces compact the sand around the pattern. Jolting produces the highest compaction at the horizontal parting line, whereas in squeezing, compaction is highest at the squeezing head (Fig. 3-7). Thus, more uniform compaction can be obtained by combining squeezing and jolting.

In vertical flaskless molding, the halves of the pattern form a vertical chamber wall against which sand is blown and compacted (Fig. 3-8). Then, the mold halves are packed horizontally, with the parting line oriented vertically and moved along a pouring conveyor. This operation is simple and eliminates the need to handle flasks, allowing for very high production rates, particularly when other aspects of the operation (such as coring and pouring) are automated.

Sandslingers fill the flask uniformly with sand under high-pressure stream. They are used to fill large flasks and are typically operated by machine. An impeller in the machine throws sand from its blades or cups at such high speeds that the machine not only places the sand but also rams it appropriately.

In impact molding, the sand is compacted by controlled explosion or instantaneous release

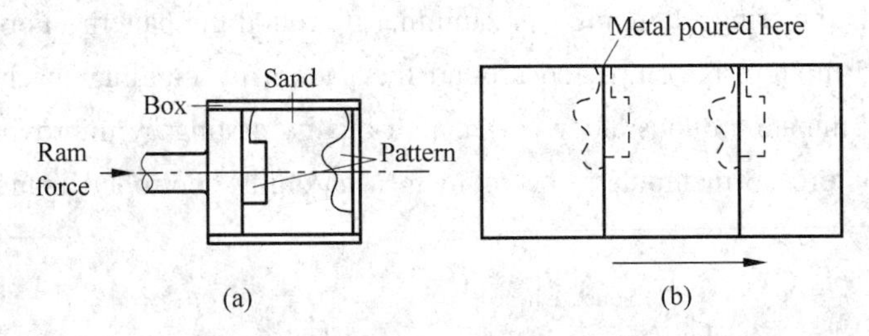

Fig. 3-8 Vertical flaskless molding

(a) sand is squeezed between two halves of the pattern; (b) assembled molds pass along an assembly line for pouring

of compressed gases. This method produces molds with uniform strength and good permeability.

In vacuum molding, also known as the "V" process, the pattern is covered tightly by a thin sheet of plastic. A flask is placed over the coated pattern and is filled with dry binderless sand. A second sheet of plastic is then placed on top of the sand, and a vacuum action hardens the sand so that the pattern can be withdrawn. Both halves of the mold are made this way and assembled.

During pouring, the mold remains under a vacuum but the casting cavity does not. When the metal has solidified, the vacuum is turned off and the sand falls away, releasing the casting. Vacuum molding produces castings with high-quality detail and dimensional accuracy. It is especially well suited for large, relatively flat castings.

3.2.6 The Sand Casting Operation

After the mold has been shaped and the cores have been placed in position, the two halves (cope and drag) are closed, clamped, and weighted down. They are weighted to prevent the separation of the mold sections under the pressure exerted when the molten metal is poured into the mold cavity.

The design of the gating system is important for proper delivery of the molten metal into the mold cavity. As described, turbulence must be minimized, air and gases must be allowed to escape by such means as vents, and proper temperature gradients must be established and maintained to minimize shrinkage and porosity. The design of risers is also important in order to supply the necessary molten metal during solidification of the casting. The pouring basin may also serve as a riser. A complete sequence of operations in sand casting is shown in Fig. 3-9. In Fig. 3-9(a), a mechanical drawing of the part is used to generate a design for the pattern. Considerations such as part shrinkage and draft must be built into the drawing. In (b) ~ (c), patterns have been mounted on plates equipped with pins for alignment. Note the presence of core prints designed to hold the core in place. In (d) ~ (e), core boxes produce core halves, which are pasted together. The cores will be used to produce the hollow area of the part shown in (a). In (f), the cope half of the mold is assembled by securing the cope pattern plate to the flask with aligning pins, and attaching inserts to form the sprue and risers. In (g), the flask is rammed with sand and the plate and inserts are removed. In (h), the drag half is produced in a similar manner, with the pattern inserted. A bottom board is placed below the drag and aligned with pins. In (i), the pattern, flask, and bottom board are inverted, and the pattern is withdrawn, leaving the appropriate imprint. In (j), the core is set in place within the drag cavity. In (k), the mold is closed by placing the cope on top of the drag and securing the assembly with pins. The flasks are then subjected to pressure to counteract buoyant forces in the liquid, which might lift the cope. In (l), after the metal solidifies, the casting is removed from the mold. In (m), the sprue and risers are cut off and recycled, and the casting is cleaned, inspected, and heat treated (when necessary).

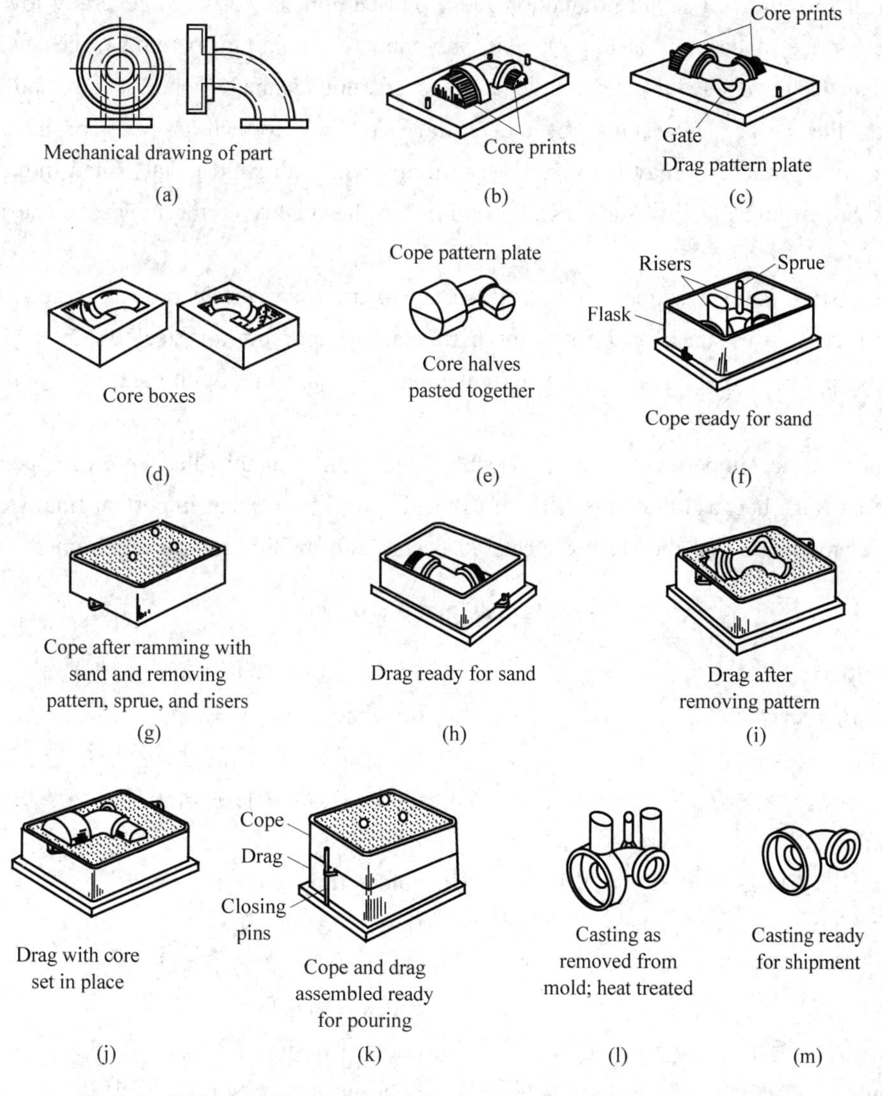

Fig. 3-9 Schematic illustration of the sequence of operations for sand casting

After solidification, the casting is shaken out of its mold, and the sand and oxide layers adhering to the casting are removed by vibration (using a shaker) or by sand blasting. Ferrous castings are also cleaned by blasting with steel shot (shot blasting) or grit. The risers and gates are cut off by oxyfuel-gas cutting, sawing, shearing, and abrasive wheels, or they are trimmed in dies. Gates and risers on steel castings are also removed with air carbon-arc or powder-injection torches. Castings may be cleaned by electrochemical means or by pickling with chemicals to remove surface oxides.

Almost all commercially-used metals can be sand cast. The surface finish obtained is largely a function of the materials used in making the mold. Dimensional accuracy is not as good as that of other casting processes. However, intricate shapes can be cast by this process, such as cast-iron engine blocks and very large propellers for ocean liners. Sand casting can be economical for relatively small production runs, and equipment costs are generally low.

The surface of castings is important in subsequent machining operations, because machinability can be adversely affected if the castings are not cleaned properly and sand particles remain on the surface. If regions of the casting have not formed properly or have formed incompletely, the defects may be repaired by filling them with weld metal. Sand-mold castings generally have rough, grainy surfaces, depending on the quality of the mold and the materials used.

The casting may subsequently be heat-treated to improve certain properties needed for its intended service use; these processes are particularly important for steel castings. Finishing operations may involve machining straightening, or forging with dies to obtain final dimensions.

Minor surface imperfections may also be filled with a metal-filled epoxy, especially for cast-iron castings because they are difficult to weld. Inspection is an important final step and is carried out to ensure that the casting meets all design and quality control requirements.

New Words and Expressions

pattern 模型
imprint 留下烙印
riser 冒口
cope 上型箱
silica sand 硅砂
bank sand 岸砂(黏土少于5%的天然砂,铸造用砂)
synthetic sand 合成砂
lake sand 湖砂
foundry 铸造,翻砂,铸造厂
permeability 渗透性
green-sand mold 湿(砂)型
cold-box mold 低温铸模
torch 割炬,焊炬,喷管,切割器
drawback 缺点,障碍
collapsibility 崩溃性,退让性
binder 黏结剂,曲面压碎圈,双动压力机外滑块
cold-setting process 冷塑(固)化过程
flask 型(砂)箱
drag 阻力,制动,牵制
cheek 耐火侧墙
pouring basin 浇口杯
pouring cup 浇口杯,外浇口
inlet 入口
blind riser 暗冒口

open riser 明冒口
hollow 孔,空穴
lettering 字体
pocket 槽
parting agent 脱模剂
one-piece pattern 整体模
split pattern 对分模,组合模
two-piece pattern 上下两件模
match-plate pattern 双面模板模
rapid prototyping 快速成形
fraction 小部分
valve 阀
interior 内部的
shakeout 抖掉,打型芯
aggregate 成套设备,机组
anchor 固定
core print 型芯座
recess 凹槽,凹处
chaplet 型芯撑
sweep 扫描,清除,弯曲
core blower 芯型吹砂机,吹芯机
compact 压紧,压实
anvil 基准面
inertial force 惯性力
conveyor 传送带,输送机
impeller 叶轮
blade 叶片,刀刃
instantaneous 瞬时的
dry binderless sand 干燥的无黏结的型砂
temperature gradient 温度梯度
porosity 多孔性,疏松度,孔隙度
oxide layer 氧化层
sand blasting 喷砂处理
grit 粗砂,研磨
abrasive wheel 砂轮,磨轮
sawing 锯,锯开
air carbon-arc torch 空气电弧焊炬
powder-injection torch 喷粉焊炬
electrochemical 电化学的
pickling 酸洗
cast-iron 铸铁
propeller 推进器
ocean liner 远洋定期客轮
drawing of the part 零件图
aligning pin 定位销
weld metal 焊接(焊缝)金属
heat-treated 对……进行热处理
straightening 校正
cast-iron casting 铸造件

3.3 Die Casting

Die casting is the art of rapidly producing accurately dimensioned parts by forcing molten metal under pressure into metal dies. The term also applies to the resultant casting. Die castings can be used economically in designs having moderate to large activity because the completed piece has a good surface, requires relatively little machining, and can be held to close tolerances. The principles of die casting follow those of good practice in any casting operation. The steel dies are permanent and should not be affected by the metal introduced into them, except for normal abrasion or wear. Die-casting dies are usually more expensive than those used in plastic or permanent molding of a part of similar size and shape.

The rapidity of operation depends upon the speed with which the metal can be forced into the die, cooled, and ejected; the casting removed; and the die prepared for the next shot.

3.3.1 The Die Casting Cycle

In the casting cycle, first the die is closed and locked. The molten metal, which is maintained by a furnace at a specified temperature, then enters the injection cylinder. Depending on the type of alloy, either a hot-chamber or cold-chamber metal-pumping system is used. These will be described later. During the injection stage of the die casting process, pressure is applied to the molten metal, which is then driven quickly through the feed system of the die while air escapes from the die through vents. The volume of metal must be large enough to overflow the die cavities and fill overflow wells. These overflow wells are designed to receive the lead portion of the molten metal, which tends to oxidize from contact with air in the cavity and also cools too rapidly from initial die contact to produce sound castings. Once the cavities are filled, pressure on the metal is increased and held for a specified dwell time during which solidification takes place. The dies are then separated, and the part extracted, often by means of automatic machine operation. The open dies are then cleaned and lubricated as needed, and the casting cycle is repeated.

Following extraction from the die, parts are often quenched and then trimmed to remove the runners, which were necessary for metal flow during mold filling. Trimming is also necessary to remove the overflow wells and any parting-line flash that is produced. Subsequently, secondary machining and surface finishing operations may be performed.

3.3.2 Die Casting Alloys

The four major types of alloys that are die-cast are zinc, aluminum, magnesium, and copper-based alloys. The die casting process was developed in the 19th century for the manufacture of lead/tin alloy parts. However, lead and tin are now very rarely die-cast because of their poor mechanical properties.

The most common die casting alloys are the aluminum alloys. They have low density, good corrosion resistance, are relatively easy to cast, and have good mechanical properties and dimensional stability. Aluminum alloys have the disadvantage of requiring the use of cold-chamber machines, which usually have longer cycle times than hot-chamber machines owing to the need for a separate ladling operation.

Zinc-based alloys are the easiest to cast. They also have high ductility and good impact strength, and therefore can be used for a wide range of products. Castings can be made with very thin walls, as well as with excellent surface smoothness, leading to ease of preparation for plating and painting. Zinc alloy castings, however, are very susceptible to corrosion and must usually be coated, adding significantly to the total cost of the component. Also, the high specific gravity of zinc alloys leads to a much higher cost per unit volume than for aluminum die casting alloys.

Zinc-aluminum (ZA) alloys contain a higher aluminum content (82. 7%) than the standard zinc alloys. Thin walls and long die lives can be obtained, similar to standard zinc alloys, but as with aluminum alloys, cold-chamber machines, which require pouring of the molten metal for each cycle, must usually be used. The single exception to this rule is ZA8 (8% Al), which has the lowest aluminum content of the zinc-aluminum family.

Magnesium alloys have very low density, a high strength-to-weight ratio, exceptional damping capacity, and excellent machinability properties.

Copper-based alloys, brass and bronze, provide the best mechanical properties of any of the die casting alloys; but they are much more expensive. Brasses have high strength and toughness, good wear resistance, and excellent corrosion resistance.

One major disadvantage of copper-based alloy casting is the short die life caused by thermal fatigue of the dies at the extremely high casting temperatures. Die life is influenced most strongly by the casting temperature of the alloys, and for that reason is greatest for zinc and shortest for copper alloys. However, this is only an approximation since casting size, wall thickness, and geometrical complexity also influence the wear and eventual breakdown of the die surface.

3.3.3 Die Casting Dies

Die casting dies consist of two major sections——the ejector die half and the cover die half——which meet at the parting line. The cavities and cores are usually machined into inserts that are fitted into each of these halves. The cover die half is secured to the stationary platen, while the ejector die half is fastened to the movable platen (see Fig. 3-10, Fig. 3-11). The cavity and matching core must be designed so that the die halves can be pulled away from the solidified casting.

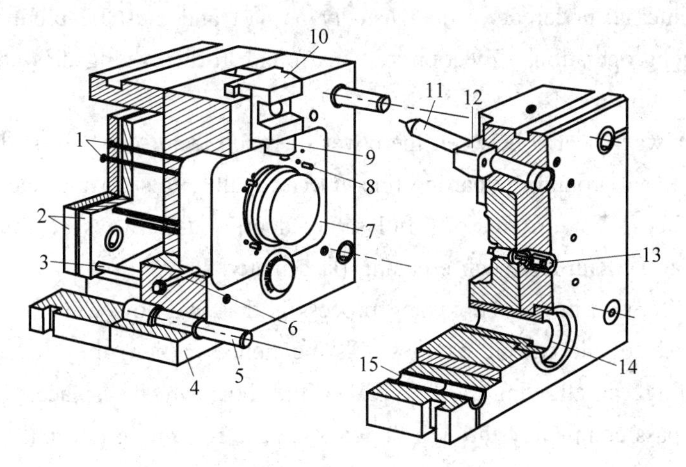

Fig. 3-10 Typical die assembly

1—Ejector pin; 2—Ejector plate; 3—Ejector return pin; 4—Base support; 5—Guide pillar; 6—Die insert waterway; 7—Die insert; 8—Fixed core; 9—Moving core; 10—Moving core holder; 11—Angle pin; 12—Core locking wedge; 13—Cascade waterway; 14—Plunger bush; 15—Guide bush

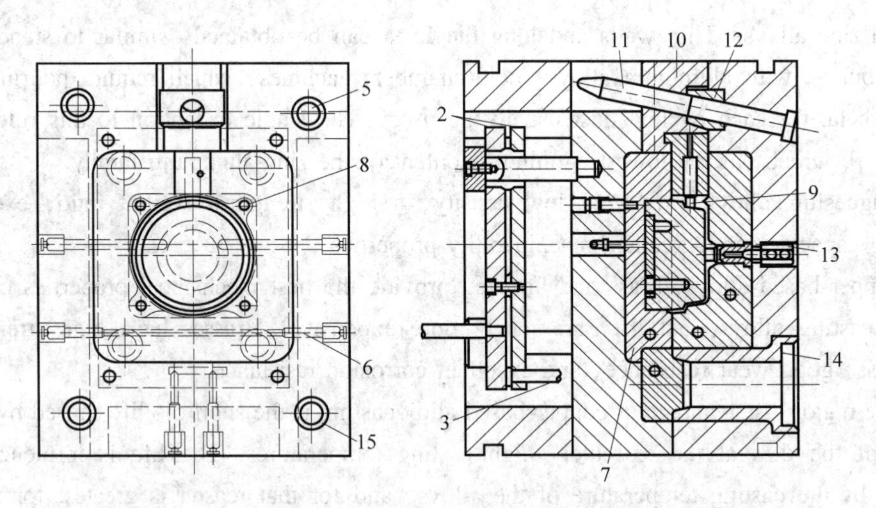

Fig. 3-11 Views illustrating structural features of die show in Fig. 3-10

The construction of die casting dies is almost identical to that of molds for injection molding. In injection molding terminology, the ejector die half comprises the core plate and ejector housing, and the cover die half comprises the cavity plate and backing support plate.

Side-pull mechanisms for casting parts with external cross-features can be found in exactly the same form in die casting dies as in plastic injection molds. However, molten die casting alloys are much less viscous than the polymer melt in injection molding and have a great tendency to flow between the contacting surfaces of the die. This phenomenon, referred to as "flashing", tends to jam mold mechanisms, which must, for this reason, be robust. The combination of flashing with the high core retraction forces due to part shrinkage makes it extremely difficult to produce satisfactory internal core mechanisms. Thus, internal screw threads or other internal undercuts cannot usually be cast and must be produced by expensive additional machining operations. Ejection systems found in die casting dies are identical to the ones found in injection molds.

"Flashing" always occurs between the cover die and ejector die halves, leading to a thin, irregular band of metal around the parting line. Occasionally, this parting line flash may escape between the die faces. For this reason, full safety doors must always be fitted to manual die casting machines to contain any such escaping flash material.

One main difference in the die casting process is that overflow wells are usually designed around the perimeter of die casting cavities. As mentioned earlier, they reduce the amount of oxides in the casting, by allowing the first part of the shot, which displaces the air through the escape vents, to pass completely through the cavity. The remaining portion of the shot and the die are then at a higher temperature, thereby reducing the chance of the metal freezing prematurely. Such premature freezing leads to the formation of surface defects called cold shuts, in which streams of metal do not weld together properly because they have partially solidified by the time they meet. Overflow wells are also needed to maintain a more uniform die temperature on small castings, by adding substantially to the mass of molten metal.

3.3.4 Die Casting Machines

1. Hot-Chamber Machines

A typical hot-chamber injection or shot system, as shown in Fig. 3-12, consists of a cylinder, a plunger, a gooseneck, and a nozzle. The injection cycle begins with the plunger in the up position. The molten metal flows from the metal-holding pot in the furnace, through the intake ports, and into the pressure cylinder. Then, with the dies closed and locked, hydraulic pressure moves the plunger down into the pressure cylinder and seals off the intake ports. The molten metal is forced through the gooseneck channel and the nozzle and into the sprue, feed system, and die cavities. The sprue is a conically expanding flow channel that passes through the cover die half from the nozzle into the feed system. The conical shape provides a smooth transition from the injection point to the feed channels and allows easy extraction from the die after solidification. After a preset dwell time for metal solidification, the hydraulic system is reversed and the plunger is pulled up. The cycle then repeats.

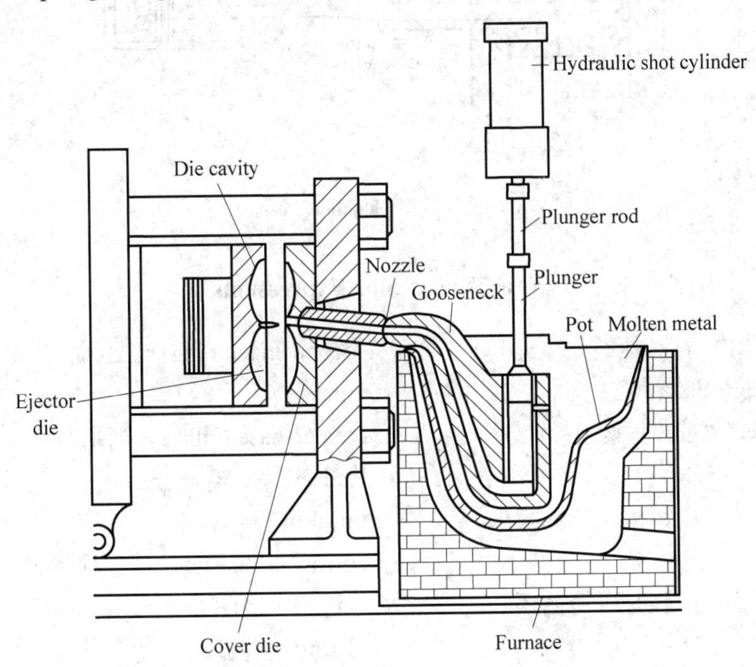

Fig. 3-12 A typical hot-chamber die-casting machine

2. Cold-Chamber Machines

A typical cold-chamber machine, as shown in Fig. 3-13, consists of a horizontal shot chamber with a pouring hole on the top, a water-cooled plunger, and a pressurized injection cylinder. The sequence of operations is as follows: when the die is closed and locked and the cylinder plunger is retracted, the molten metal is ladled into the shot chamber through the pouring hole. In order to tightly pack the metal in the cavity, the volume of metal poured into the chamber is greater than the combined volume of the cavity, the feed system, and the overflow wells. The injection cylinder is then energized, moving the plunger through the chamber, thereby forcing the molten metal into the die cavity. After the metal has solidified, the die opens and the plunger moves back to its original position. As the die opens, the excess metal at the end of the injection cylinder, called the biscuit, is forced out of the cylinder because it is attached to the casting. Material in the biscuit is required during the die casting cycle in order to maintain liquid metal pressure on the casting while it solidifies and shrinks.

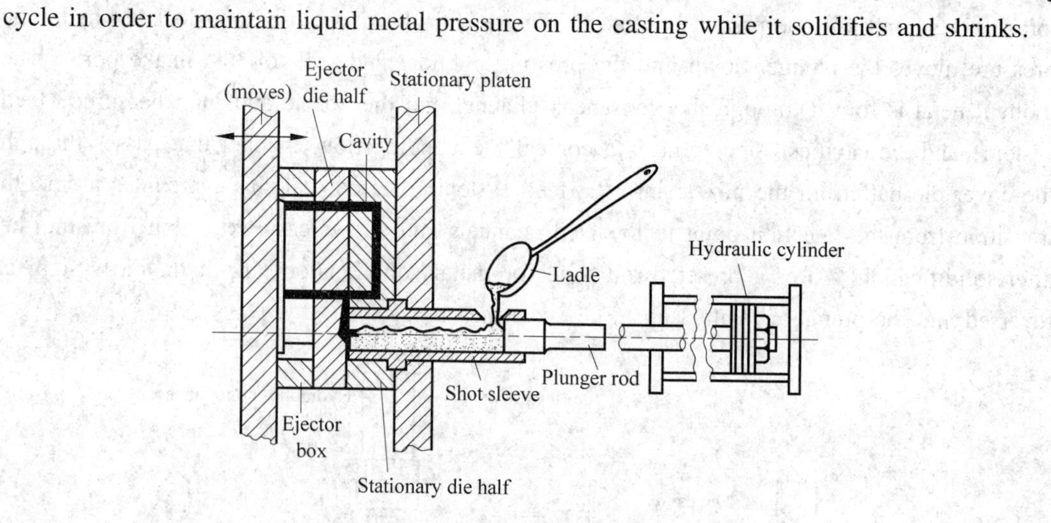

Fig. 3-13 A typical cold-chamber die-casting machine

New Words and Expressions

die-casting 压力铸造,压铸
resultant casting 最终铸造
close tolerance 小公差,紧公差
abrasion 磨损
die-casting mold 压铸模
shot 注射
casting cycle 压铸周期,压铸循环
furnace 熔炉
hot-chamber 热压室
cold-chamber 冷压室
metal-pumping system 金属浇注系统
overflow well 溢流井
lead portion 料头
oxidize 氧化
dwell time 保压时间
quench 淬火
second machining 二次加工
zinc 锌
copper-based alloy 铜基合金
lead alloy 铅合金
tin alloy 锡合金
corrosion resistance 抗腐蚀性
ladling 浇注
plating 电镀
corrosion 侵蚀
zinc-aluminum alloy 锌铝合金
magnesium alloy 镁合金
strength-to-weight ratio 强度重量比
machinability 机械加工性,切削性
brass 黄铜
bronze 青铜
thermal fatigue 热疲劳

cover die　定模
stationary pattern　固定台,固定板
ejector return pin　顶杆回程销
base support　底板,底座
guide pillar　导柱
die insert waterway　压模嵌件水路
die insert　压模嵌件
fixed core　固定型芯
moving core　活动型芯
moving core holder　活动型芯架
angle pin　斜导柱
core locking wedge　型芯锁楔
cascade waterway　串联水路
plunger bush　柱塞衬套
side-pull mechanism　侧拉机构
robust　坚固的,健康的
retraction　回复
perimeter　周长,周边
premature　早期的
cold shuts　冷寒,冷疤
gooseneck　鹅颈管,流道

Chapter 4

CNC Machining

4.1 Introduction to CNC

4.1.1 Concept of NC and CNC

Numerical control (NC) is a form of programmable automation in which the mechanical actions of a machine tool or other equipment are controlled by a program containing coded alphanumeric data. The alphanumerical data represent relative positions between a workhead and a workpart as well as other instructions needed to operate the machine. The workhead is a cutting tool or other processing apparatus, and the workpart is the object being processed. When the current job is completed, the program of instructions can be changed to process a new job. The capability to change the program makes NC suitable for low and medium production. It is much easier to write new programs than to make major alterations of the processing equipment.

Numerically controlled (NC) machine tools were developed to fulfill the contour machining requirements of complex aircraft parts and forming dies. The first-generation numerically controlled units used digital electronic circuits and did not contain any actual central processing unit, thereby they were called NC or hardwired NC machine tools. In 1970s, computer numerically controlled (CNC) machine tools were developed with minicomputers used as control units. With the advances in electronics and computer technology, current CNC systems employed several high-performance microprocessors and programmable logical controllers that work in a parallel and coordinated fashion. Current CNC systems allow simultaneous servo position and velocity control of the axis, monitoring of controller and machine tool performance, online part programming with graphical assistance, in-process cutting process monitoring, and in-process part gauging for completely unmanned machining operations. Manufacturers offer most of these features as options. Today, virtually all the new machine control units are based on computer technology hence, when we refer to NC in chapter and elsewhere, we mean CNC.

4.1.2 Basic Component of NC Machine Tools

The control system of a numerically controlled machine tool can handle many tasks commonly done by the operator of a conventional machine. For this, the numerical control system must "know" when and in what sequence it should issue commands to change tools, at what speeds and feeds the machine tool should operate, and how to work a part to the required size. The system gains the ability to perform the control functions through the numerical input information that is the control program, also called the part program.

The work process of NC is shown in Fig. 4-1. The part programmer should study the part drawing and the process chart and then prepare the control program on a standard form in the specified format. It contains all the necessary control information. A computer-assisted NC part program for NC machining method is also available, in which the computer considerably facilitates the work of the programmer and generate a set of NC instructions. Next the part program is transferred into the control computer the wide accepted method is that the worker types the part program into the computer from the keyboard of the computer numerical control front panel. The computer converts each command into the signal that the servo-drive unit needs. The servo-drive unit drives the machine tool to manufacture the finished part.

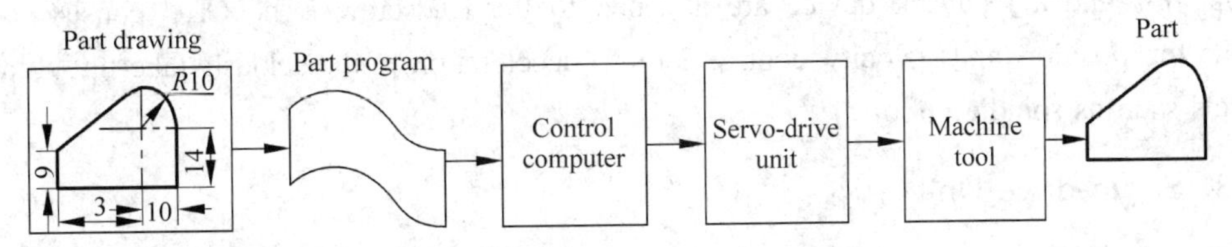

Fig. 4-1 The work process of NC

A typical NC machine tool has five fundamental units. (1) the input media, (2) the machine control unit, (3) the servo-drive unit, (4) the feedback transducer, and (5) the mechanical machine tool unit. The general relationship among the five components is illustrated in Fig. 4-2.

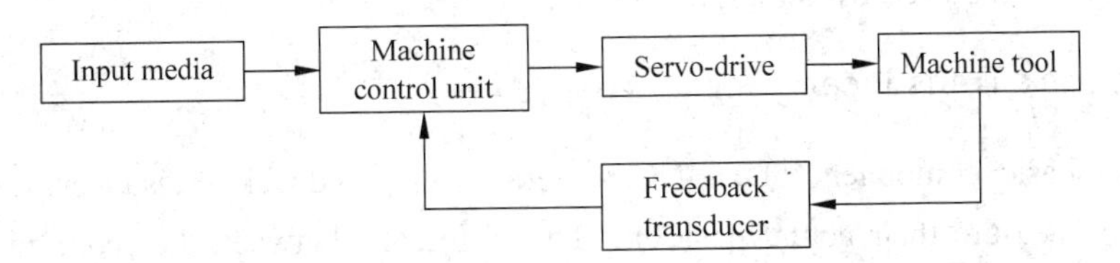

Fig. 4-2 Basic components of a CNC machine tool

1. Input Media

The input media contains the program of instructions, it is the detailed step-by-step commands that direct the actions of the machine tool; the program of instructions is called a part program. The individual commands refer to positions of a cutting tool relative to the worktable on which the workpart is fixtured. Additional instructions are usually included, such as spindle speed, feed rate, cutting tool selection, and other functions. The program is coded on a suitable medium for submission to the machine control unit. For many years, the common medium was 1-inch wide punched tape, using a standard format that could be interpreted by the machine control unit. Today, punched tape has largely been replaced by newer storage technologies in modern machine shops. These technologies include magnetic tape, diskette, and electronic transfer of part programs from a computer.

2. Machine Control Unit

The machine control unit (MCU) is a microcomputer that stores the program and executes the commands into actions by the machine tool. The MCU consists of two main units: the Data Processing Unit (DPU) and the Control Loops Unit (CLU). The DPU software includes control system software, calculation algorithms, translation software that converts the part program into a usable format for the MCU, interpolation algorithm to achieve smooth motion of the cutter, editing of part program. The DPU processes the data from the part program and provides it to the CLU which operates the drives attached to the machine leadscrews and receives feedback signals on the actual position and velocity of each one of the axes. A driver (dc motor) and a feedback device are attached to the leadscrew. The CLU consists of the circuits for position and velocity control loops, deceleration and backlash take-up, function controls such as spindle on/off.

3. Servo-drive Unit

The drives in machine tools are classified as spindle and feed drive mechanisms. Spindle and feed drive motors and their servo-amplifiers are the components of the servo-drive unit. The MCU processes the data and generates discrete numerical position commands for each feed drive and velocity command for the spindle drive. The numerical commands are converted into signal voltage by the MUC unit and sent to servo-amplifiers, which process and amplify them to the high voltage levels required by the drive motors.

4. Feedback Transducer

The forth basic component of an NC system is the feedback transducer. As the drives move, sensors measure their actual position. The difference between the required position and the actual position is detected by comparison circuit and the action is taken, within the servo, to minimize this difference.

5. Machine Tool

The machine tool performs useful work. It accomplishes the processing steps to transform the starting workpiece into a completed part. Its operations are directed by the MCU, which in turn is driven by instructions contained in the part program. In the most common example of NC, machine tool consists of the worktable and spindle.

New Words and Expressions

NC (numerical control) 数控
CNC (computer numerical control) 数控机床
programmable automation 可编程自动控制
alphanumeric data 文字数字数据
workhead 工作夹具
workpart 工件
computer numerically controlled 计算机数字控制的
machine tool 机床
microprocessor 微处理器
programmable logical controller 可编程逻辑控制器
parallel and coordinated fashion 并行协调方式
servo control 伺服控制
velocity control 速率控制
graphical assistance 图形帮助
in-process 进程中
part program 零/部件加工程序
part programmer 零件程序设计员
process chart 进程图
panel 面板
machine leadscrew 机床丝杠
feedback signal 反馈信号
deceleration 减速
backlash 后座
servo-drive unit 伺服驱动单元
input media 输入设备
machine control unit 机床控制单元
feedback transducer 反馈传感器
step-by-step command 步进命令
fixtured 固定的
submission 提交
punched tape 打孔带
magnetic tape 磁带
diskette 磁盘
electronic transfer 电子传递
system software 系统软件
calculation algorithm 计算法则
transition software 转换软件
feed drive mechanism 进给驱动机构
servo-amplifier 伺服放大器
signal voltage 信号电压
drive motor 驱动机构,传动装置
comparison circuit 比较电路

4.2 The CNC Program

4.2.1 Axes

In programming, it is necessary to label the direction that the tool travels to give commands to the machine.

In machines such as the milling machine that move the table instead of the tool, it must be understood that the direction the tool is cutting is what is important, not the table movement. Therefore, on a milling machine, as the tool cuts from right to left or left to right this is called the *X* axis and is the longitudinal movement (Fig. 4-3). As the table moves in or out toward the column and away from it (crosswise), this is the *Y* axis. When the tool moves vertically with the spindle, it is called the Z axis.

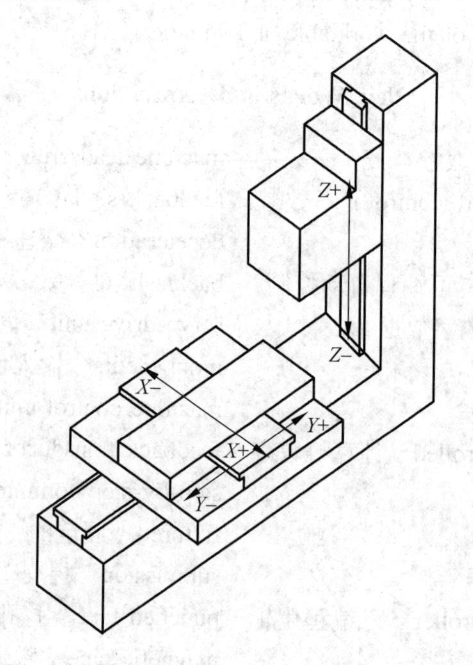

Fig. 4-3 The axes on a vertical milling machine

When turning on a lathe, the Z axis is movement of the carriage toward or away from the spindle and the X axis is the movement of the carriage toward or away from the operator (crosswise). There is no Y axis on a lathe unless an attachment may be put on it.

Note that the movement of the tool represents a quadrant (Fig. 4-4). The programmer controls the movement of the tool using the coordinate system.

Once understood, programming arcs or straight lines is easy. In an arc, for example, the points where the arc starts are the beginning, as plotted out on the coordinates, then the point where the arc ends are plotted. The radius is given in distance. Now the arc can be plotted, or machined very easily.

To determine what to call the movement of the axis when going from one direction or the other on a milling machine, the following is used:

(1) The X axis is positive (+) when the tool is cutting to the right and negative (−) when the tool is cutting to the left (Fig. 4-5).

(2) The Y axis is positive (+) when the tool is cutting away from the column and negative (−) when the tool is cutting toward the column.

(3) The Z axis is positive (+) when the tool is moving up or away from the workpiece and negative (−) when it is plunging in or moving into the workpiece.

On the lathe, the following directions are used:

(1) The X axis is positive (+) when the tool is cutting toward the operator or the carriage is moving away from the operator. It is negative (−) when the tool is cutting away from the operator or the carriage is moving toward the operator.

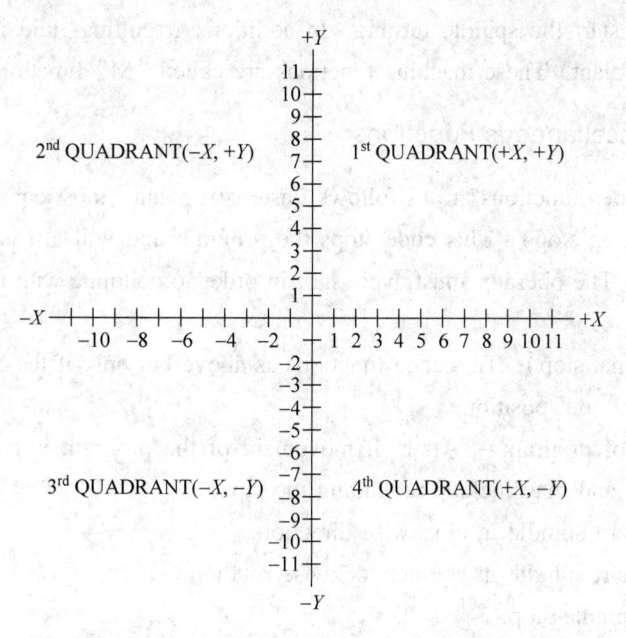

Fig. 4-4 Four quadrants in the Cartesian coordinate system

(2) The Z axis is positive (+) when the cutting tool is moving to the operator's right or away from the spindle. The Z axis is negative (−) when the cutting tool is cutting toward the spindle or moving to the operator's left.

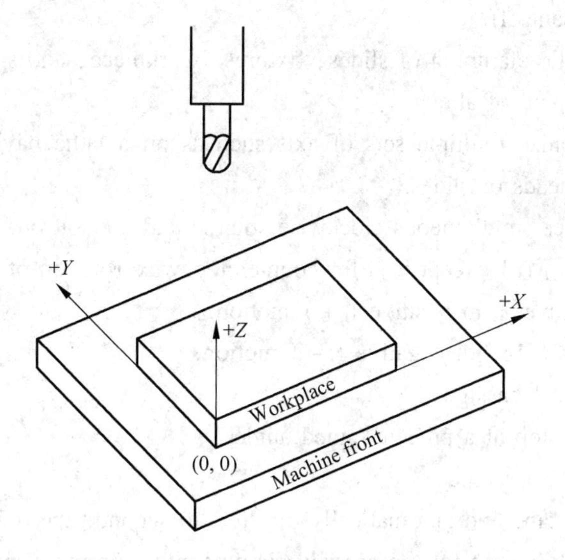

Fig. 4-5 Vertical Z axis in relation to X and Y axes, the workpiece, and the table

4.2.2 Programming

Once the axes are understood along with mathematics and blueprint reading, programming a part can readily be accomplished. There are some very basic machine codes that are used and reference to them is necessary. First are the miscellaneous functions used to start the machine, select the tool, and start the spindle turning. In addition, if cutting fluid is used it starts and stops the flow of coolant. These machine functions are called "M" functions.

4.2.2.1 Miscellaneous Functions

The miscellaneous functions are as follows (asterisks mean "most commonly used").

* M00 (program stop). This code stops the program and will also stop the spindle and coolant if activated. The operator must cycle start in order to continue with the remainder of the program.

* M01 (optional stop). This code functions as above but only if the control unit optional stop selector is in the "on" position.

* M02 (end of program). After all movement of the program is completed, this code will stop everything and may include rewinding the tape.

* M03. * Start spindle in clockwise direction.

* M04. * Start spindle in counterclockwise rotation.

* M05. * Spindle stop.

* M06 (tool change). * This code stops the spindle and coolant and retracts the spindle to allow a tool change.

* M07. * Turn flood coolant on.

* M08. * Turn mist coolant on.

* M09. * Coolant off.

* M10. Automatic clamping of slides, fixtures, workpiece, and spindle.

* M11. Unclamping of above.

* M12. Synchronize multiple sets of axis such as on a lathe having four axis and two independent operated heads or slides.

* M13. Combines simultaneous clockwise spindle and coolant on.

* M14. Same as M13 except it is for counterclockwise rotation of spindle.

* M15. Rapid traverse in positive (+) motion.

* M16. Same as M15 but negative (-) motion.

* M17 and M18. Unassigned.

* M19. Spindle stop at a predetermined angle.

* M20 to M29. Unassigned.

* M30. Rewind tape and automatically transfers to second tape if in the system.

* M31. Interlock bypass for temporarily circumventing normally provided interlock.

* M32 to M39. Unassigned.

* M40 to M46. Signal gear changes are required at the machine; otherwise unassigned.

* M47. Continues program execution from the start of the program unless inhibited by an interlock signal.

* M48. Cancels M49

* M49. Deactivates a manual spindle or feed override and returns to the original program

value.

* M50 to M57 Unassigned.

* M58. Cancels M59.

* M59. Holds RPM constant at program value.

* M60 to M99. Unassigned.

4.2.2.2 Preparatory Functions

Next are the "G" codes, known as the preparatory functions. These codes determine such modes of operation of the system as drill cycles, dwells, rapid traverse, cutter compensation, and many others. The following G codes are as follows (asterisks mean "most commonly used").

* G00. * Rapid traverse with point-to-point positioning.

* G01. * For describing linear interpolation blocks and reserved for contouring.

* G02. * Clockwise circular interpolation.

* G03. * Counter clockwise circular interpolation.

* G04. * Time delay (dwell).

* G05 to G07. Unassigned. Can be used by machine tool builder or systems builder.

* G08. Acceleration code. Causes the machine to accelerate at a smooth rate.

* G09. Deceleration rate.

* G10 to G12. Not used with CNC. Used with hardwire systems.

* G13 to G16. Used to direct the control system to operate on a particular set of axes.

* G17 to G19. To select a coordinate plane for such functions as circular interpolation or cutter compensation.

* G20 to G32. Unassigned. Can be used by the machine manufacturer.

* G33 to G35. * Used for CNC lathes for thread cutting. G33 is for constant lead on threads, G34 is for increasing lead, and G35 is for decreasing lead on threads.

* G36 to G39. Unassigned.

* G40. * Terminate cutter compensation.

* G41. * Cutter compensation, when the cutter is on the left side of the work surface looking in the direction of the cutter motion.

* G42. * Cutter compensation, when the cutter is on the right side of the work surface.

* G43. * Cutter compensation, to an inside corner.

* G44. * Cutter compensation, to an outside corner.

* G45 to G49. Unassigned.

* G50 to G59. Reserved for adaptive control.

* G60 to G69. Unassigned.

* G70. * Inch programming.

* G71. * Metric programming.

* G72. Three-dimensional circular interpolation, clockwise.

* G73. Three-dimensional circular interpolation, counterclockwise.
* G74. Cancel multi-quadrant circular interpolation.
* G75. Multi-quadrant circular interpolation.
* G76 to G79. Unassigned.
* G80. Cancel cycle.
* G81. * Drill cycle.
* G82. * Drill with dwell.
* G83. * Intermittent or deep hole drilling.
* G84. * Tap cycle.
* G85 to G89. Boring cycles.
* G90. * Absolute programming.
* G91. * Incremental programming.
* G92. Preload values such as axis-position registers.
* G93. Inverse-time feed rate.
* G94. * Inches or millimeters feed rate.
* G95. * Inches or millimeters per revolution feed rate.
* G97. * Spindle speed in RPM.
* G98 and G99. Unassigned.

4.2.2.3 Other Address Characters

The other address characters used are as follows (asterisks mean "most commonly used").

* A. Angular dimensions about the *X* axis.
* B. Angular dimensions about the *Y* axis.
* C. Angular dimensions about the *Z* axis.
* D. Can be used for an angular dimension around a special axis, for a third feed function, or for tool offset.
* E. Same as D but for a second feed function.
* F. * Feed rate.
* H. Unassigned.
* I, J, K. * Used with circular interpolation. You will see these most frequently.
* L. Not used.
* M. * Miscellaneous functions.
* N. * Sequence number.
* O. Used in place of the customary sequence number word address N.
* P. A third rapid traverse code parallel to the *X* axis.
* Q. Second rapid traverses code parallel to the *Y* axis.
* R. First rapid traverse code parallel to the *Z* axis.
* T. * Tool change number.
* U. Secondary motion parallel to the *X* axis.

* V. Secondary motion parallel to the *Y* axis.
* W. Secondary motion parallel to the *Z* axis.

4.2.2.4 Absolute Programming and Incremental Programming

These are the two methods used in programming a part.

Absolute programming is using a reference or common point and positioning the cutter in all positions with reference to that starting point. The advantage of absolute positioning is that an error in positioning that occurs in one place will not progressively recur.

Incremental positioning is positioning the cutter from the last or previous position. If an error is made in incremental positioning, then all subsequent positions will be in error. This is not to say either method is better for they both are used, depending on what the program requires.

In programming, these can be interchanged throughout the program. It all depends on the type of program desired.

4.2.2.5 Sample Program

In this particular example, we are milling around the outside of a workpiece contour (shown in Fig. 4-6). Notice that we are using a one-inch diameter end mill for machining the contour and we are programming the very center of the end mill.

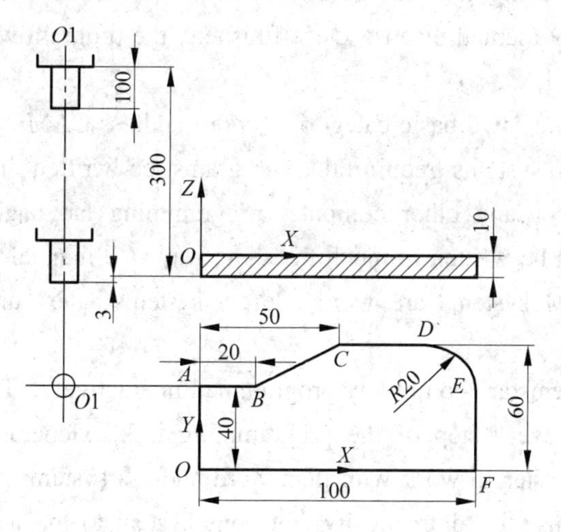

Fig. 4-6 The sketch of workpiece

O0001 (Program number)

N10 G21 G90 (The metric system, absolute mode)

N20 G92 X-50 Y40 Z300 (Select coordinate system)

N30 G00 Z-3 G43 H01 M03 S300 (Instate tool length compensation, rapid tool down to work surface, and start spindle at 300r/min)

N40 G41 X-5 D01 (Instate tool radius compensation, and close with point *A*)

N50 G01 X20 Y40 F100 (Machine in straight motion to point *B*, the feed is 100 mm/min)

N60 X50 Y60 (Machine in straight motion to point *C*)

N70 X80 (Machine in straight motion to point *D*)

N80 G02 X100 Y40 R20 (Circular motion to point *E*)

N90 G01 Y0 (Machine in straight motion to point *F*)

N100 X0 (Machine in straight motion to point *O*)

N110 Y40 (Machine in straight motion to point *A*)

N120 G00 X-50 Y40 (Rapid to the machine's reference point in *XY* plane)

N130 G00 G40 Z300 (Cancel tool compensation, and rapid to the machine's reference point in *Z*)

N140 M02 (End of program)

4.2.3 CAM System Programming

CAM systems allow CNC programming to be accomplished at a much higher level than manual programming and are becoming very popular. Generally speaking, a CAM system helps the programmer in three major areas. It keeps the programmer from having to do math calculations, makes it easy to program different kinds of machines with the same basic language, and helps with certain basic machining practice functions.

With a CAM system, the programmer will have a computer to help with the preparation of the CNC program. The computer will actually generate the G-code level program much like a CNC program created by manual means. Once finished, the program will be transferred directly to the CNC machine tool.

CAM systems fall into two basic categories, word address CAM systems and graphic CAM systems. Word address systems require that programs be written in a language similar to BASIC, C Language, or any other computer programming language. These CAM systems require that the program be written in much the same way as a manual program. While some of the most powerful CAM systems are word address systems, they also tend to be the more difficult to use.

Graphic CAM systems are commonly programmed interactively. The programmer will have visual feedback during every step of the programming task. Generally speaking, this makes graphic CAM systems easier to work with than word address systems.

While CAM systems vary dramatically from one system to the next, there are three basic steps that remain remarkably similar among most of them. First, the programmer must give some general information. Second, workpiece geometry must be defined and trimmed to match the workpiece shape. Third, the machining operations must be defined.

1) General Information

Information required of the programmer in this step includes documentation information like part name, part number, date, and program file name. The programmer may also be required to set up the graphic display size for scaling purposes. The workpiece material and rough stock shape may also be required.

2) Define and Trim Geometry

Using a series of geometry definition methods, the programmer will describe the shape of the workpiece. With graphic CAM systems, the programmer will generally be shown each geometric element as it is described. The programmer will have the ability to select from a series of definition methods, choosing the one that makes it the easiest to define the workpiece shape.

Once geometry is defined, most CAM systems require that the geometry be trimmed to match the actual shape of the workpiece to be machined. Lines that run off the screen in both directions must be trimmed to form line segments. Circles must be trimmed to form radii.

3) Bypassing the Geometry Creation

Keep in mind that most CAM systems allow geometry defined within CAD(computer-aided design) systems to be imported to the CAM system. This is especially helpful with very complicated parts, keeping the CAM system programmer from having to duplicate the effort of creating geometry.

New Words and Expressions

Axes 轴线
milling machine 铣床
column 立柱
crosswise 成十字状的,交叉的
lathe 车床,机床
carriage 滑动架
quadrant 象限,四分仪
coordinate system 坐标系统
positive 正的,正数的
negative 负的,负数的
workplace 工作面,工作场所
machine front 机器前端
mathematics 数学运算,数学应用
miscellaneous function 辅助功能
asterisks 星号,加星号的
coolant 冷却剂,冷却液
tape 线,带
clockwise 顺时针方向的,右旋的
counterclockwise 逆时针方向的,左旋的
flood coolant 水冷或液体冷却
mist coolant 喷雾冷却
fixture 夹具,固定装置
synchronize 同步,同时发生
unassigned 未定义,未分配,未赋值
predetermined 预定的,预先确定的
rewind 倒回,重绕,回卷
interlock 互锁
circumventing 规避,绕行
gear 齿轮,传动装置
deactivate 撤销,停用
feed override 进给倍率
preparatory function 准备功能
drill cycle 钻孔
dwell 暂停
rapid traverse 快速移动,快速行程
cutter compensation 刀具补偿
describing 沿……运行,形成……形状
interpolation 插入,插补
acceleration 加速
hardwire 硬件,固定线路
thread cutting 螺纹切削
lead 螺距
terminate 结束,终止
adaptive control 自适应控制
tap 攻螺纹
boring 镗孔,镗削

absolute programming 绝对值编程
incremental programming 增量值编程
preload 预载,预加负载
register 寄存器
inverse-time 时间倒数
angular 角度,用角度测量
traverse 横贯,横穿,横跨
recur 重复,递归
interchange 互换,相互交换
instate 任命
word address 字符地址,字符代码
graphic 图解的,坐标式的
geometric element 几何元素
radii 半径

4.3 Machining Centers

Machining center has evolved from individual machines which, with the aid of man, performed individual processes to machines capable of performing many processes.

In 1968, a NC machine was marketed which could automatically change tools so that many different processes could be done in one machine. Such a machine became known as a "machining center"——a machine that can perform a variety of processes and change tools automatically while under programmable control.

The study of machining centers begins with the history of numerical control (NC).

Computer and numerical control is used on a wide variety of machines. These range from single spindle drilling machines, which often have only two-axis control, to machining centers, which can do drilling, boring, milling, tapping, and so forth with four axis control. A machining center can automatically select and change as many as 32 preset tools. The table can move left/right or in/out and the spin die can move up/down or in/out, with positioning accuracy in the range of 0.003 in. in 40 in. of travel. The machine has automatic tool change and automatic work transfer so that workpiece can be loaded/unloaded while the machining is in process.

The concept of automatic tool changing has been extended to NCN lathes. The tools are held on a rotating tool magazine and a gantry-type tool changer is used to change the tools. Each magazine holds one type of cutting tool. The versatility is being increased by combining both rotary-work and rotary-tool operations——turning and milling——in a single machine. Tools are changed in six seconds or less. It is also common to provide two or more worktables, permitting work to be set up while machining is done on the workpiece in the machine, with tables being interchanged automatically. Consequently, the productivity of such machines can be very high, the chip producing time often approaching 50% of the total.

The main parts of CNC machining center is a bed, saddle, column, table, servo motors, ball screws, spindle, tool changer, and the machine control unit.

1. Bed

The bed is usually made of high-quality cast iron which provides for a rigid machine capable of performing heavy-duty machining and maintaining high precision. Hardened and ground ways are mounted to the bed to provide rigid support for all linear axes.

2. Saddle

The saddle, which is mounted on the hardened and ground bedways, provides the machining center with the X axis linear movement.

3. Column

The column, which is mounted to the saddle, is designed with high torsion strength to prevent distortion and deflection during machining. The column provides the machining center with the Y axis linear movement.

4. Table

The table, which is mounted on the bed, provides the machining center with the *Z* axis linear movement.

5. Servo System

The servo system, which consists of servo drive motors, ball screws, and position feedback encoders, provides fast, accurate movement and positioning of the *XYZ* axis slides. The feedback encoders mounted on the ends of the ball screws form a closed-loop system which maintains consistent high-positioning unidirectional repeatability of +0.0001 in. (0.003 mm).

6. Spindle

The spindle, which is programmable in 1 r/min increments, has a speed range of from 20 to 6,000 r/min. The spindle can be of a fixed position (horizontal) type, or can be a tilting/contouring spindle which provides for an additional axis.

7. Tool Changers

There are basically two types of tool changers, the vertical tool changer and the horizontal tool changer. The tool changer is capable of storing a number of preset tools which can be automatically called for use by the part program. Tool changers are usually bidirectional, which allows for the shortest travel distance to randomly access a tool. The actual tool change time is usually only 3 to 5 s.

8. MCU

The MCU allows the operator to perform a variety of operations such as programming, machining, diagnostics, tool and machine monitoring, etc. MCUs vary according to manufacturers' specifications; new MCUs are becoming more sophisticated, making machine tools more reliable and the entire machining operations less dependent on human skills.

Two new trends are observed in the development of machining centers. One is the growing interest in smaller, more compact machining centers and the other is the emphasis on extended shift or even unmanned operations. Modern machining centers have contributed significantly to improved productivity in many companies. They have eliminated the time lost in moving workpieces from machine to machine and the time needed for workpiece loading and unloading for separate operations. In addition, they have minimized the time lost in changing tools, carrying out gauging operations, and aligning workpieces on the machine.

New Words and Expressions

machining center 加工中心
preset tool 预调刀具
spin die 旋转模头
rotating tool magazine 旋转式刀具库
gantry-type tool changer 龙门刀具变换器
rotary-work 回转工件
rotary-tool 回转刀具
producing time 生产期
bed 机床
saddle 鞍,滑板,座板
table 工作台
ball screw 滚珠丝杠
tool changer 刀库
heavy-duty machining 重型机械
bedways 床身导轨
torsion strength 转矩
encoder 编码器
unidirectional repeatability 单向重复性
bidirectional 双向的
unmanned operation 无人操作
gauging operation 测量

4.4 Automation of Manufacturing

4.4.1 Introduction

Automation is the technology by which a process or procedure is accomplished without human assistance. It is implemented by using a program of instructions combined with a control system that executes the instructions. To automate a process, power is required, both to drive the process itself and to operate the program and control system. Although automation can be applied in a wide variety of areas, it is most closely associated with the manufacturing industries. It was in the context of manufacturing that the term was originally coined by an engineering manager at Ford Motor Company in 1946 to describe the variety of automatic transfer devices and feed mechanisms that had been installed in Ford's production plants. It is ironic that nearly all modern applications of automation are controlled by computer technologies that were not available in 1946.

Automated manufacturing systems operate in the factory on the physical product. They perform operations such as processing, assembly, inspection, or material handling, in some cases accomplishing more than one of these operations in the same system. They are called automated because they perform their operations with a reduced level of human participation compared with the corresponding manual process. In some highly automated systems, there is virtually no human participation. Examples of automated manufacturing systems include:

(1) Automated machine tools that process parts;

(2) Transfer lines that perform a series of machining operations;

(3) Manufacturing systems that use industrial robots to perform processing or assembly operations;

(4) Automatic material handing and storage systems to integrate manufacturing operations;

(5) Automatic inspection systems for quality control.

Automated manufacturing systems can be classified into three basic types: (1) fixed automation, (2) programmable automation, and (3) flexible automation.

Fixed automation is a system in which the sequence of processing (or assembly) operations is fixed by the equipment configuration. Each of the operations in the sequence is usually simple, involving perhaps a plain linear or rotating motion or an uncomplicated combination of the two. It is the integration and coordination of many such operations into one piece of equipment that makes the system complex. Typical features of fixed automation are:

(1) High initial investment for custom-engineered equipment;

(2) High productivity rates;

(3) Relatively inflexible in accommodating product variety.

The economic justification for fixed automation is found in products that are produced in very large quantities and at high production rates. The high initial cost is its character compared with alternative methods of production. Examples of fixed automation include machine transfer lines and automated assembly machines.

In programmable automation, the production equipment is designed with the capability to change the sequence of operations to accommodate different product configurations. The operation sequence is controlled with a program, which is a set of instructions coded so that they can be read and interrupted by the system. New programs can be prepared and entered into the equipment to produce new products. Programmable automated production systems are used in low- and medium-volume production. The parts or products are typically made in batches. To produce each new batch of a different product, the system must be reprogrammed with the set of machine instructions that correspond to the new product. The physical setup of machine must also be changed: tools must be loaded, fixtures must be attached to the machine table, and the required machine settings must be entered. Examples of programmable automation include numerically controlled machine tools, industrial robots, and programmable logic controllers.

Flexible automation is an extension of programmable automation. A flexible automated system is capable of producing a variety of parts (or products) with virtually no time lost for changeovers from one part style to the next. There is no lost production time while reprogramming the system and altering the physical setup (tooling, fixtures, machine settings). Consequently, the system can produce various combinations and schedules of part or products instead of requiring that they be made in batches. What makes flexible automation possible is that the differences between parts processed by the system are not significant. It is a case of soft variety, so that the amount of changeover required between styles is minimal. Examples of flexible automation are flexible manufacturing systems for performing machining operations that date back to the late 1960s.

4.4.2 Flexible Manufacturing System

In the modern manufacturing setting, flexibility is an important characteristic. It means that a manufacturing system is versatile and adaptable, while also capable of handling relatively high production runs. A flexible manufacturing system is versatile in that it can produce a variety of parts. It is adaptable because it can be quickly modified to produce a completely different line of parts.

A flexible manufacturing system (FMS) is a highly automated TG (group technology) machine cell, consisting of a group of processing workstations (usually CNC machine tools), interconnected by an automated material handing and storage system, and controlled by a distributed computer system. The reason the FMS is called flexible is that it is capable of processing a variety of different part styles simultaneously at the various workstations, and the mix of patterns. The FMS is most suited for the mid-variety, mid-volume production range.

An FMS relies on the principles of group technology. Group technology is a manufacturing philosophy in which similar parts are identified and grouped together to take advantage of their similarities in design and production, similar parts are arranged into part families, where each part family possesses similar design and manufacturing characteristics. No manufacturing system can be completely flexible. There are limits to the range of parts or productions that can be made in an FMS. Accordingly, an FMS is designed to produce parts (or products) within a defined range of styles, sizes, and processes. In other words, an FMS is capable of producing a single part family or a limited range of part families.

An FMS must possess three capabilities: (1) the ability to identify and distinguish among the different part or product styles processed by the system, (2) quick changeover of operating instructions, and (3) quick changeover of physical setup.

4.4.3 Computer Integrated Manufacturing System

Computer integrated manufacturing (CIM) is the term used to describe the modern approach to manufacturing. Although CIM encompasses many of the other advanced manufacturing technologies such as computer numerical control (CNC), computer-aided design/computer-aided manufacturing (CAD/CAM), robotics, and just-in-time delivery (JIT), it is more than a new technology or a new concept. Computer integrated manufacturing is an entirely new approach to manufacturing, a new way of doing business.

To understand CIM, it is necessary to begin with a comparison of modern and traditional manufacturing. Modern manufacturing encompasses all of the activities and processes necessary to convert raw materials into finished products, deliver them to the market, and support them in the field. These activities include the following:

(1) Identifying a need for a product;

(2) Designing a product to meet the needs;

(3) Obtaining the raw materials needed to produce the product;

(4) Applying appropriate processes to transform the raw materials into finished products;

(5) Transporting product to the market;

(6) Marinating the product to ensure proper performance in the field.

This broad, modern view of manufacturing can be compared with the more limited traditional view that focused almost entirely on the conversion processes. The old approach excluded such critical pre-conversion elements as market analysis research, development, and design, as well as such after-conversion elements as product delivery and product maintenance. In other words, in the old approach to manufacturing, only those processes that took place on the shop floor were considered manufacturing. This traditional approach of separating the overall concept into numerous stand-alone specialized elements was not fundamentally changed with the advent of automation.

With CIM, not only are the various elements automated, but also the islands of automation are all linked together or integrated. Integration means that a system can provide complete and instantaneous sharing of information. In modern manufacturing, Integration is accomplished by computers. CIM, then, is the total integration of all components involved in converting raw materials into finished products and getting the products to the market.

New Words and Expressions

automation 自动操作
automatic transfer device 自动转换装置
feed mechanism 进给机构,进料机构
transfer line 转换线
automated machine tool 自动化机床
manufacturing system 制造系统
industrial robot 工业机器人
automatic material handing system 自动送料系统
fixed automation 固定自动机
flexible automation 柔性自动化
equipment configuration 设备装置
plain linear motion 水平直线运动
rotating motion 旋转运动
initial investment 最初投资
custom-engineered equipment 工程设备
productivity rate 生产率
economic justification 经济因素
operation sequence 工序
low- and medium-volume production 中小量生产
batch 一批
changeover 转换
physical setup 物理安装
production run 生产过程
group technology 成组工艺,成组技术
machine cell 机加工单元
processing workstation 处理站
flexible manufacturing system (FMS) 柔性制造系统
manufacturing philosophy 制造原理
part family 零件族

computer integrated manufacturing 计算机一体化制造
computer-aided design (CAD) 计算机辅助设计
computer-aided manufacturing (CAM) 计算机辅助制造
robotics 机器人技术
just-in-time delivery 即时运输
marinate 浸泡
integration 集成,一体化

Chapter 5

Mold Special Machining

5.1 Electrical Discharge Machining

Electrical discharge machining (EDM) is a machining method primarily used for hard metals or those that would be impossible to machine with traditional techniques. EDM is sometimes called "spark machining", because it removes metal by producing a rapid series of repetitive electrical discharges. A shaped graphite or copper electrode is used to make a cavity that is the mirror image of the electrode, without direct contact with the workpiece. Sparks travel through a dielectric fluid, typically a light oil, at a controlled distance. EDM can machine cavities with thin walls and fine features, achieve difficult part geometry, produce burr-free parts, and is insensitive to workpiece hardness. Surfaces of EDM parts have a characteristic cratered appearance, with all craters being the same size.

5.1.1 Principle of EDM

It has been recognized for many years that a powerful spark will cause pitting or erosion of the metal at both the anode(+) and cathode(−), e.g., automobile battery terminals, loose plug points, etc.. This process is utilized in EDM. The EDM process involves a controlled erosion of electrically conductive materials by the initiation of rapid and repetitive spark discharges between the tool and workpiece separated by a small gap of about 0.01 to 0.5mm. This spark gap is either flooded or immersed in a dielectric fluid. The controlled pulsing of the direct current between the tool and the work produces the spark discharge.

Initially, the gap between the tool and the workpiece, which consists of the dielectric fluid, is not conductive. However, the dielectric fluid in the gap is ionised under pulsed application of DC as shown in Fig. 5-1, thus enables the spark discharge to pass between the tool and the both tool and the workpiece melts, partially vaporizes and partially ionises the metal in a thin surface layer. Due to the inertia of the surrounding fluid, the pressure within the spark becomes quite large and may possibly assist in "blasting" the molten material from the surface leaving a fairly flat and shallow crater. The amount of material removed per spark depends upon the electrical energy expended per spark and the period over which it is expended.

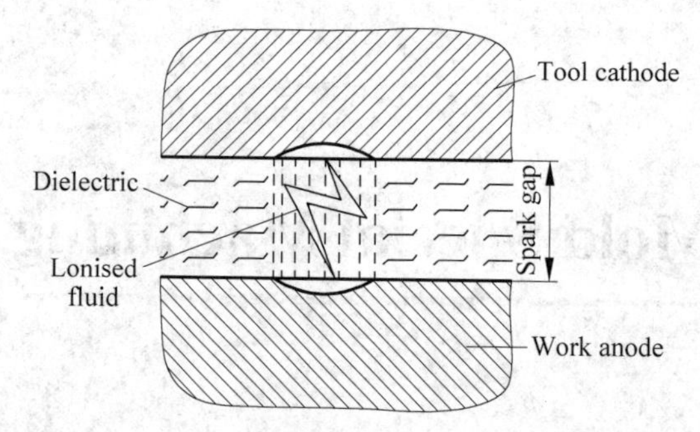

Fig. 5-1 Schematic of the arc formation in EDM process

Thus, the sequence of events in EDM can be summarized as follows:

1. With the application of voltage, and electric field builds up between the two electrodes at the position of least resistance. The ionization leads to the breakdown of the dielectric which results in the drop of the voltage and the beginning of flow of the current.

2. Electrons and ions migrate to anode and cathode respectively at very high current density. A column of vapour begins to form and the localized melting of work commences. The discharge channel continues to expand along with a substantial increase of temperature and pressure.

3. When the power is switched off, the current drops; no further heat is generated and the discharge column collapses. A portion of molten metal evaporates explosively and/or is ejected away from the electrode surface. With the sudden drop in temperature, the remaining molten and vaporized metal solidifies. A tiny crater is thus generated at the surface.

4. The residual debris is flushed away along with products of decomposition of dielectric fluid. The application of voltage initiates the next pulse and the cycle of events.

At any given time, only one spark will be made between the tool and workpiece at the shortest path as shown in Fig. 5-2. As a result of this spark, some volume of metal is removed from both the tool and the workpiece. Then the spark will move to the next closest distance as shown in Fig. 5-2. This process continues till the required material is removed from the workpiece.

The temperature of the arc may reach about 10,000℃. The vapour of the metal would be quenched by the dielectric medium when the arc is terminated by the electric pulse and thus the wear debris is always spherical in nature. The wear debris would be carried away by the dielectric fluid, which is in continuous circulation. The same process as described above would be continued a number of times per second with each pulse removing a small wear particle from the workpiece, thereby causing the material to take the shape of the electrode. The arc will always be struck at a point between the workpiece which is closed from the tool (electrode),

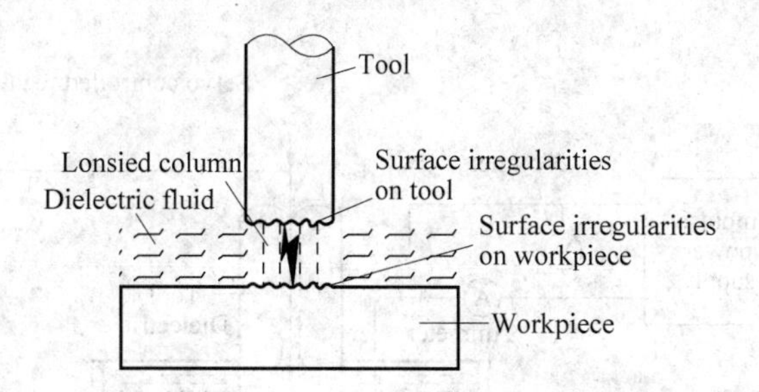

Fig. 5-2 Schematic diagram of the arc moving to the next smallest distance between the tool and the workpiece in the EDM process

hence the complimentary tool surface will be reproduced in the workpiece as shown in Fig. 5-3.

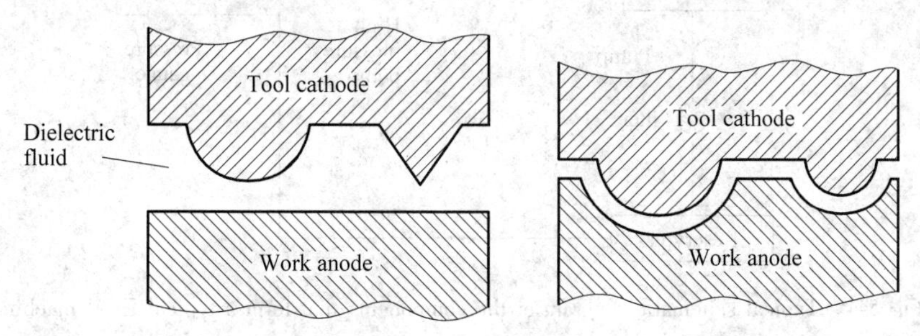

Fig. 5-3 Typical surface generation in EDM process

(a) Initial shape of electrode and workpiece; (b) Final complimentary shapes of electrode and workpiece after machining

A typical schematic diagram of the various elements present in a commercial EDM machine is shown in Fig. 5-4. The main power unit consists of the required controlled pulse generator with the DC power to supply the power pulses. The pulse frequency as well as the on-and-off time of the pulses can be very accurately controlled by using electronic controllers.

The movements of the EDM machine are determined by a computer numerical control (CNC) controller on the machine that has been programmed by the tool engineering department. The program to cut this ratchet was loaded into the EDM machine's controller memory and the program run to produce the part to great accuracy.

5.1.2 Wire EDM

There are two primary EDM methods: ram EDM and wire EDM. The primary difference between the two involves the electrode that is used to perform the machining. In a typical ram EDM application, a graphite electrode is machined with traditional tools. The now specially-shaped electrode is connected to the power source, attached to a ram, and slowly fed into the workpiece. The entire machining operation is usually performed while submerged in a fluid bath. In wire EDM, a very thin wire serves as the electrode. Special brass wires are typically used; the wire is slowly fed through the material and the electrical discharges actually cut the workpiece. Wire EDM is usually performed in a bath of water.

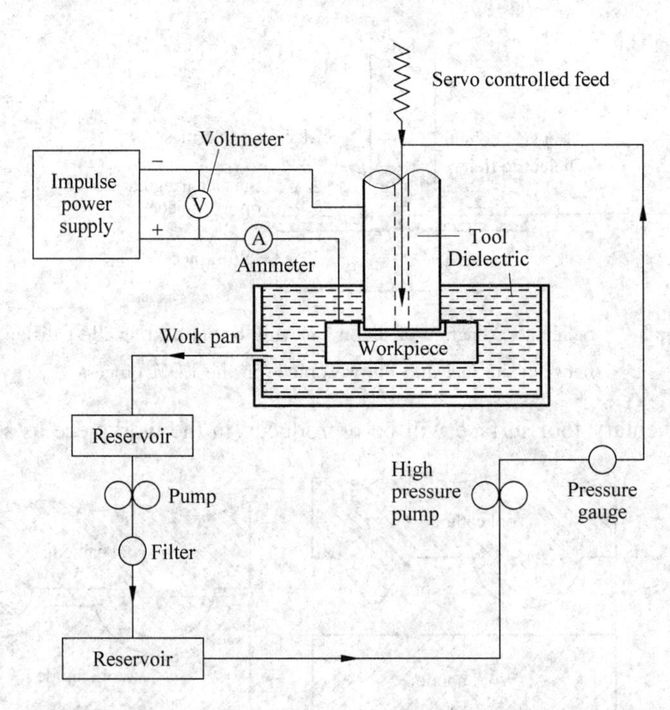

Fig. 5-4 Typical schematic diagram of the components that form a typical EDM machine

1. The basic principle of Wire EDM

The basic principle of Wire EDM(WEDM) is that the continuously moving thin metal wire is taken as tool electrode and the pulsating current is linked to the metal wire and the workpiece, and then with the impulse spark discharging between the metal wire and the workpiece, the metal is made to be melted or to be gasified. During the relative movement of the electrode wire and workpiece, the workpiece is cut and shaped. WEDM is also called wire cut. A typical schematic diagram of a wire EDM operation is shown in Fig. 5-5.

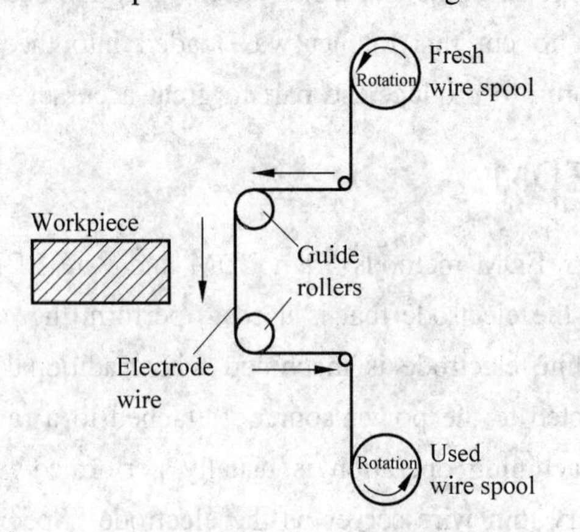

Fig. 5-5 Principle of wire EDM process

2. The component of wire EDM machine tool

Wire EDM machine tool (shown in Fig. 5-6) is mainly composed of machine itself, impulse power, working liquid cycle system, NC system and machine accessories and so on.

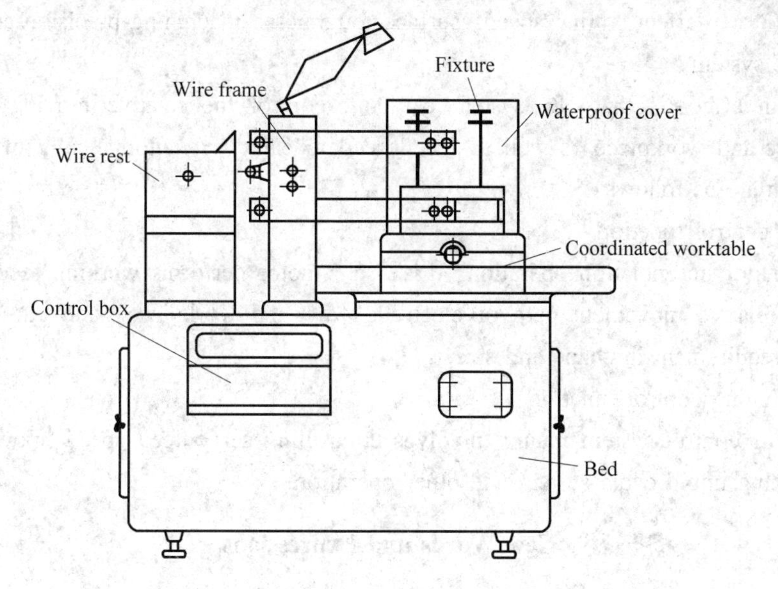

Fig. 5-6 The component of wire EDM machine tool

(1) Machine itself

1) Bed

Bed is a support of the coordinate worktable, storing wire device and wire rest. It should have enough strength and rigidity. Power and working liquid box may be mounted inside the bed.

2) Coordinated worktable

The coordinated worktable is used to mount workpieces and performs the predetermined relative movement with respect to electrode wire according to control requirement. It involves carrier, way and drive device.

3) Wire rest

The effect is that the wire rest can support and guide the electrode wire by means of two guide pulleys on the wire rest and can make some parts of the electrode wire and worktable keep certain angle to perform taper cutting. Double coordinate gang wire rest has added U. V. two drive motors to carry out taper cutting by program control.

(2) Impulse power

The effect is that the impulse power change power frequency alternating current into single impulse current at a certain frequency in order to provide electric energy which workpiece and electrode wire requires when they discharging to corrode metal. Its character directly affects machining speed, surface quality, machining precision and loss in electrode wire.

There is a variety of impulse power. They are transistor rectangle wave impulse power, high-frequency grouped impulse power, parallel capacitance-type impulse power and low-loss impulse power and so on.

(3) Working liquid cycle system

In the machining, working liquid plays roles of insulation, flushing, chip-removal, cooling, and it can affect cutting speed, surface roughness, machining precision and so on.

(4) NC system

The main effect is that NC system can auto-control the relative moving path of the electrode wire and workpiece as well as feed speed to carry out automatically machining. Its main function are as follows:

1) Path control function

By means of interpolation operation, drive step motor performs working feed, accurately controls the relative movement path of electrode wire and workpiece, and cuts parts which accord with requirement in shape and size.

2) Machining control function

Machining control system mainly involves controlling servo feed speed, power unit wire device, working liquid cycle system and other operation.

New Words and Expressions

Electrical Discharge Machining (EDM) 电火花加工
spark machining 电火花加工
repetitive 反复的
graphite 石墨
electrode 电极
dielectric fluid 介电液体,电介质
burr-free 无毛刺
crater 有坑,形成坑
pitting 点蚀
anode 阳极
cathode 阴极
plug point 插座
direct current 直流电
ionize 电离
pulsed 使产生脉冲
vaporize 气化,蒸发
blasting 爆炸,爆破
ionization 电离
electron 电子
ion 离子
switched off 关闭
debris 碎片,残骸
spherical 球形的,圆的
pulse generator 脉冲发生器
quenched 淬火
ratchet 棘轮
voltmeter 电压计,电压表
ammeter 电流计,安培表
reservoir 蓄水池,储水池
pump 泵
filter 过滤器
pressure gauge 压力计
servo controlled feed 伺服控制进给
pulsating current 脉冲电流
specially-shaped 异形的
gasified 气化
wire spool 丝卷
electrode wire 电极丝
guide roller 导轮
taper cutting 锥度切割
impulse power 脉冲电源

coordinate worktable 坐标工作台
wire rest 运丝装置
guide pulleys 引导滑轮
transistor 晶体管
capacitance-type 电容式
insulation 绝缘
flushing 清洗
chip-removal 排屑
automatically machining 自动加工
interpolation operation 插补运算
servo feed 伺服进给
control box 操纵盒
wire frame 丝架
waterproof cover 防水罩

5.2 Electrochemical Machining

Electrochemical machining (ECM) is a method of removing metal by an electrochemical process and its basis is the phenomenon of electrolysis. Metal removal is achieved by electrochemical dissolution of an anodically polarized workpiece which is one part of an electrolytic cell in ECM. It is normally used for mass production and is used for working extremely hard materials or materials that are difficult to machine using conventional methods. Its use is limited to electrically conductive materials. ECM can cut small or odd-shaped angles, intricate contours or cavities in hard and exotic metals, such as titanium aluminides, inconel, waspaloy, and high nickel, cobalt, and rhenium alloys. Both external and internal geometries can be machined.

ECM is often characterized as "reverse electroplating", in that it removes material instead of adding it. It is similar in concept to electrical discharge machining (EDM) in that a high current is passed between an electrode and the part, through an electrolytic material removal process having a negatively charged electrode (cathode), a conductive fluid (electrolyte), and a conductive workpiece (anode); however, in ECM there is no tool wear. The ECM cutting tool is guided along the desired path close to the work but without touching the piece. Unlike EDM, however, no sparks are created. High metal removal rates are possible with ECM, with no thermal or mechanical stresses being transferred to the part, and mirror surface finishes can be achieved.

The ECM process is most widely used to produce complicated shapes such as turbine blades with good surface finish in difficult to machine materials. It is also widely and effectively used as a deburring process. In deburring, ECM removes metal projections left from the machining process, and also dulls sharp edges. This process is fast and often more convenient than the conventional methods of deburring by hand or nontraditional machining processes.

5.2.1 Working Principle of ECM

Electrochemical machining is founded on the principles outlined in Fig. 5-7, the workpiece and tool are the anode and cathode, respectively, of an electrolytic cell, and a constant potential difference, usually at about 10 V, is applied across them. A suitable electrolyte, for example, aqueous sodium chloride (table salt) solution, is chosen so that the cathode shape remains unchanged during electrolysis. The electrolyte is also pumped at a rate 3 to 30 meter/second, through the gap between the electrodes to remove the products of machining and to diminish unwanted effects, such as those that arise with cathodic gas generation and electrical heating. The rate at which metal is then removed from the anode is approximately in inverse proportion to the distance between the electrodes. As machining proceeds, and with the simultaneous movement of the cathode at a typical rate, for example, 0.02 millimeter/second toward the anode, the gap width along the electrode length will gradually tend to a steady-state value. Under these conditions, a shape, roughly complementary to that of the cathode, will be reproduced on the anode. A typical gap width then should be about 0.4 millimeter.

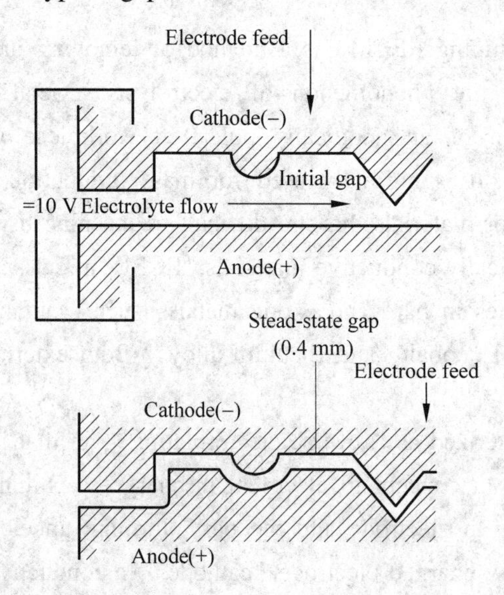

Fig. 5-7 Working principles of the ECM

5.2.2 ECM Equipment

A schematic view of the various elements present in an ECM is shown in Fig. 5-8. Generally the commercial ECM machines are characterized by the large sizes in view of the high power supplies involved. Typical power supplies used may range from 500 A to 4000 A.

The tool-to-work gap needs to be maintained at a very small value of the order of 0.25 mm for satisfactory metal removal rates. The electrolyte needs to be pumped through this gap at high pressures ranging from 0.70 to 3.00 MPa. This introduces a large amount of load on the machine, because of the large working areas involved. For example, if the working area is 800 cm^2, with an electrolyte pressure of 1.0 MPa, the resulting load on the tool will be 80000N. Hence the machine structure will have to be made rigid to withstand such forces.

The electrolyte consists of the metal debris removed from the anode, which will have to be filtered before it is re-pumped into the system. Also a large amount of heat is generated during the electrolysis, which heats up the electrolyte, and hence it needs to be cooled. The electrical

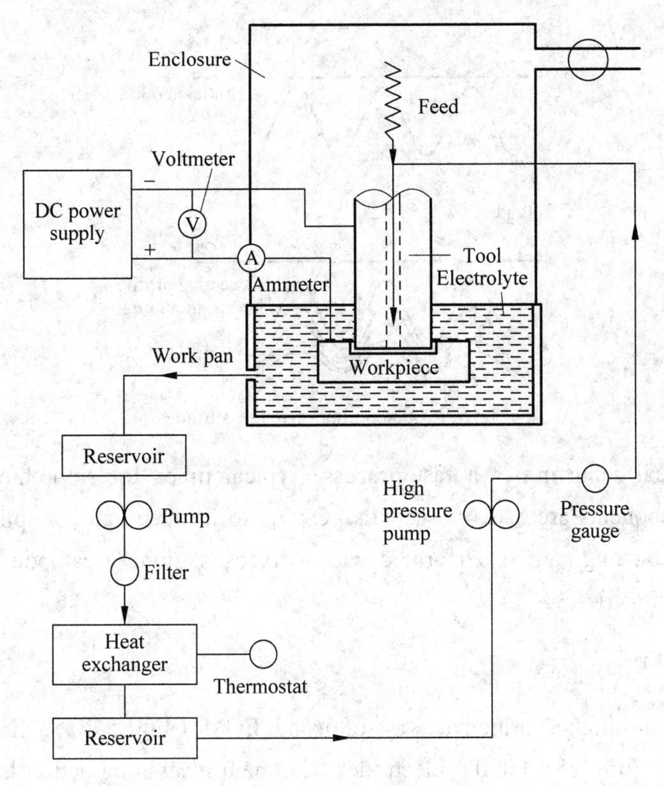

Fig. 5-8 Schematic view of the various elements present in a commercial ECM machine

conductivity of the electrolyte changes with temperature. A constant equilibrium gap needs to be maintained during the ECM operation. A servo drive is provided on the tool axis for this purpose.

The electrolytes used in the ECM process are corrosive in nature and hence proper care needs to be taken to see that all the materials that come in contact with the electrolyte be made of stainless steel, plastic or other materials to withstand the corrosion. Similarly, provision needs to be made to safely and continuously exhaust the hydrogen gas generated during the process with explosion-proof blowers.

5.2.3 Applications of ECM

1. Smoothing of rough surfaces

Deburring, or smoothing of surfaces (Fig. 5-9), is the simplest and a common use of ECM. A plane-faced cathode tool is placed opposite a workpiece that has an irregular surface. The current densities at the peaks of the surface irregularities are higher than those in the valleys. The former are removed preferentially and the workpiece becomes smooth, admittedly at the expense of stock metal (which is still machined from the valleys of the irregularities, even though at a lower rate). Electrochemical smoothing is the only type of ECM in which the final anode shape may match precisely that of the cathode tool.

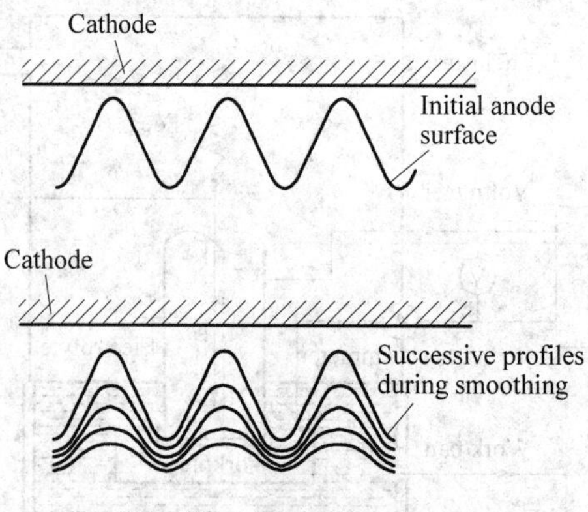

Fig. 5-9 Smoothing of rough surfaces

Electrochemical deburring is a fast process; typical times for smoothing the surfaces of manufactured components are 5 to 30 seconds. Owing to its speed and simplicity of operation, electrochemical deburring can be performed with a fixed, stationary cathode tool. The process is used in many industries.

2. Hole drilling

Hole drilling is another principal way of using ECM (Fig. 5-10). The cathode-tool is usually made in the form of a tubular electrode. The main machining action is carried out in the gap formed between the leading edge of the drill tool and the base of the hole in the workpiece. ECM also proceeds laterally between the side walls of the tool and component, where the current density is lower than at the leading edge of the advancing tool. Since the lateral gap width becomes progressively larger than that at the leading edge, the side-ECM rate is lower. The overall effect of the side-ECM is to increase the diameter of the hole produced. The distance between the side wall of the workpiece and the central axis of the cathode tool is larger than the external radius of the cathode. This difference is known as the "overcut". The amount of overcut can be reduced by several methods. A common procedure involves the insulation of the external walls of the tool (Fig. 5-10), which inhibits side current flow. Another practice lies in the choice of an electrolyte such as sodium nitrate, which has the greatest current

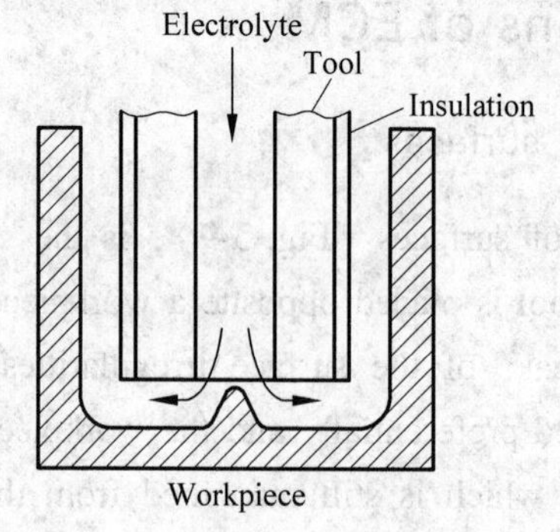

Fig. 5-10 Electrochemical hole drilling

efficiency at the highest current densities. In hole drilling, these high current densities occur between the leading edge of the drill and the base of the workpiece. If another electrolyte such as sodium chloride were used, the overcut could be much greater. The current efficiency for sodium chloride remains steady at almost 100% for a wide range of current densities. Thus, even in the side gap, metal removal proceeds at a rate that is mainly determined by the current density, in accordance with Faraday's law.

Holes with diameters of 0.05 to 75 millimeter have been achieved with ECM. For holes of 0.5 to 1.0 millimeter diameter, depths of up to 110 millimeter have been produced. Drilling by ECM is not restricted to round holes; the shape of the workpiece is determined by that of the tool electrode.

3. Full-form shaping

Full-form shaping utilizes a constant gap across the entire workpiece and the tool is moved mechanically at a fixed rate toward the workpiece in order to produce the type of shape used for the production of compressor and turbine blades. In this procedure, current densities as high as 100 A/cm^2 are used, and across the entire face of the workpiece, the current density remains high.

Electrolyte flow plays an even more influential role in full-form shaping than in drilling and smoothing of surfaces. The entire large cross-sectional area of the workpiece has to be supplied by the electrolyte as it flows between electrodes. The larger areas of electrodes involved mean that comparatively higher pumping pressures and volumetric flow rates are needed.

4. Electrochemical grinding

The main feature of electrochemical grinding (ECG) is the use of a grinding wheel in which an insulating abrasive, such as diamond particles, is set in a conducting material, as shown in Fig. 5-11. This wheel becomes the cathode tool. The non-conducting particles act as a spacer between the wheel and workpiece, providing a constant inter-electrode gap, through which electrolyte is flushed.

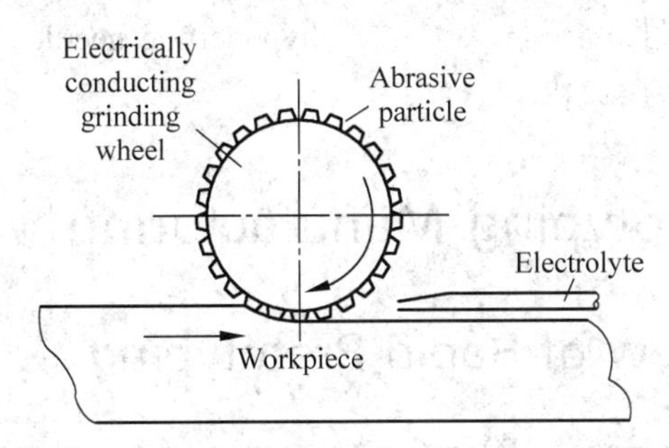

Fig. 5-11 Schematic diagram of an electrochemical grinding operation

Accuracies achieved by ECG are usually about 0.125 millimeter. A drawback of ECG is the loss of accuracy when inside corners are ground. Because of the electric field effects, radii better than 0.25 ~ 0.375 millimeter can seldom be achieved.

A wide application of electrochemical grinding is the production of tungsten carbide cutting tools. ECG is also useful in the grinding of fragile parts such as hypodermic needles.

New Words and Expressions

electrochemical machining 电解加工
electrolysis 电解作用
electrochemical dissolution 电化学溶解
anodically polarize 阳极极化
electrolytic cell 电解槽
odd-shaped 奇形的
exotic 特殊
titanium aluminide 钛铝合金
inconel 铬镍铁合金
nickel 镍
waspaloy 镍基合金
cobalt 钴
rhenium 铼
electroplating 电镀
electrolyte 电解质,电解液
conductive fluid 导电流体
turbine blade 涡轮叶片
deburring 去毛刺
metal projection 金属凸出部分
dull 钝的,不锋利的
aqueous sodium chloride 氯化钠溶液
table salt 食盐
servo drive 伺服驱动
corrosive 腐蚀的,腐蚀性的
explosion-proof blower 防爆风机
complementary 互补的
apparatus 仪器
conductivity 电导率
hole drilling 钻孔
tubular electrode 管状电极
reversal 反向,颠倒
current density 电流密度
overall effect 整体效果
central axis 中心轴
overcut 过切
side-current flow 侧电流
sodium nitrate 硝酸钠
sodium chloride 氯化钠
Faraday's law 法拉第定律
full-form shaping 整体造型
compressor 压缩机
electrochemical grinding 电解磨削
grinding wheel 砂轮
insulating abrasive 绝缘磨料
diamond particle 金刚石颗粒
spacer 垫片
inter-electrode 两极之间的
tungsten carbide 碳化钨
hypodermic needle 注射针

5.3 Rapid Prototyping Manufacturing

5.3.1 Overview of Rapid Prototyping

The term rapid prototyping (RP) refers to a class of technologies that can automatically construct physical models from Computer-Aided Design (CAD) data. These "three dimensional printers" allow designers to quickly create tangible prototypes of their designs, rather than just two-dimensional pictures. Such models have numerous uses. They make excellent visual aids for communicating ideas with co-workers or customers. In addition, prototypes can be used for design testing. For example, an aerospace engineer might mount a model airfoil in a wind tunnel to measure lift and drag forces. Designers have always utilized prototypes; RP allows them to be made faster and less expensively.

In addition to prototypes, RP techniques can also be used to make tooling (referred to as rapid tooling) and even production-quality parts (rapid manufacturing). For small production runs and complicated objects, rapid prototyping is often the best manufacturing process available. Of course, "rapid" is a relative term. Most prototypes require from three to seventy-two hours to build, depending on the size and complexity of the object. This may seem slow, but it is much faster than the weeks or months required to make a prototype by traditional means such as machining. These dramatic time savings allow manufacturers to bring products to market faster and more cheaply. In 1994, Pratt & Whitney achieved "an order of magnitude [cost] reduction [and] ... time savings of 70 to 90 percent" by incorporating rapid prototyping into their investment casting process.

At least six different rapid prototyping techniques are commercially available, each with unique strengths. Because RP technologies are being increasingly used in non-prototyping applications, the techniques are often collectively referred to as solid free-form fabrication, computer automated manufacturing, or layered manufacturing. The latter term is particularly descriptive of the manufacturing process used by all commercial techniques. A software package "slices" the CAD model into a number of thin (0.1mm) layers, which are then built up one atop another. Rapid prototyping is an "additive" process, combining layer of paper, wax, or plastic to create a solid object. In contrast, most machining processes (milling, drilling, grinding, etc.) are "subtractive" processes that remove material from a solid block. RP's additive nature allows it to create objects with complicated internal features that cannot be manufactured by other means.

Of course, rapid prototyping is not perfect. Part volume is generally limited to 0.125 m^3 or less, depending on the RP machine. Metal prototypes are difficult to make, though this should change in the near future. For metal parts, large production runs, or simple objects, conventional manufacturing techniques are usually more economical. These limitations aside, rapid prototyping is a remarkable technology that is revolutionizing the manufacturing process.

5.3.2 The Basic Process

Although several rapid prototyping techniques exist, all employ the same basic five-step process. The steps are:

(1) Create a CAD model of the design;

(2) Convert the CAD model to .STL format;

(3) Slice the STL file into thin cross-sectional layers;

(4) Construct the model one layer atop another;

(5) Clean and finish the model.

CAD Model Creation: First, the object to be built is modeled using a Computer-Aided Design (CAD) software package. Solid modelers, such as Pro/ENGINEER, tend to represent 3D objects more accurately than wire-frame modelers such as AutoCAD, and will therefore yield better results. The designer can use a pre-existing CAD file or may wish to create one expressly for prototyping purposes. This process is identical for all of the RP build techniques.

Conversion to STL Format: The various CAD packages use a number of different algorithms to represent solid objects. To establish consistency, the STL (stereolithography, the first RP technique) format has been adopted as the standard of the rapid prototyping industry. The second step, therefore, is to convert the CAD file into STL format. This format represents a three-dimensional surface as an assembly of planar triangles, "like the facets of a cut jewel." The file contains the coordinates of the vertices and the direction of the outward normal of each triangle. Because STL files use planar elements, they cannot represent curved surfaces exactly. Increasing the number of triangles improves the approximation, but at the cost of bigger file size. Large, complicated files require more time to pre-process and build, so the designer must balance accuracy with manageability to produce a useful STL file. Since the .STL format is universal, this process is identical for all of the RP build techniques.

Slice the STL File: In third step, a pre-processing program prepares the STL file to be built. Several programs are available, most allow the user to adjust the size, location and orientation of the model. Build orientation is important for several reasons. First, properties of rapid prototypes vary from one coordinate direction to another. For example, prototypes are usually weaker and less accurate in the z (vertical) direction than in x-y plane. In addition, part orientation partially determines the amount of time required to build the model. Placing the shortest dimension in the z direction reduces the number of layers, thereby shortening build time. The pre-processing software slices the STL model into a number of layers from 0.01 mm to 0.7 mm thick, depending on the build temperature. The program may also generate an auxiliary structure to support the model during the build. Supports are useful for delicate features such as overhangs, internal cavities, and thin-walled sections. Each RP machine manufacturer supplies their own proprietary pre-processing software.

Layer by Layer Construction: The fourth step is the actual construction of the part. Using one of several techniques (described in the next section) RP machines build one layer at a time from polymers, paper, or powdered metal. Most machines are fairly autonomous, needing little human intervention.

Clean and Finish: The final step is post-processing. This involves removing the prototype from the machine and detaching any supports. Some photosensitive materials need to be fully cured before use. Prototypes may also require minor cleaning and surface treatment. Sanding, sealing, and/or painting the model will improve its appearance and durability.

5.3.3 RP Techniques

In recent years, several types of rapid prototyping have emerged. The technologies developed include stereolithography (SL) , selective laser sintering (SLS) , fused deposition modeling(FDM) , laminated object modeling(LOM).

1. Stereolithography

Patented in 1986, stereolithography started the rapid prototyping revolution. The technique builds three-dimentional models from liquid photosensitive polymers that solidify when exposed to ultraviolet light. As shown in Fig. 5-12 below, the model is built upon a platform situated just below the surface in a vat of liquid epoxy or acrylate resin. A low-power highly focused UV laser traces out the first layer, solidifying the model's cross section while leaving excess areas liquid.

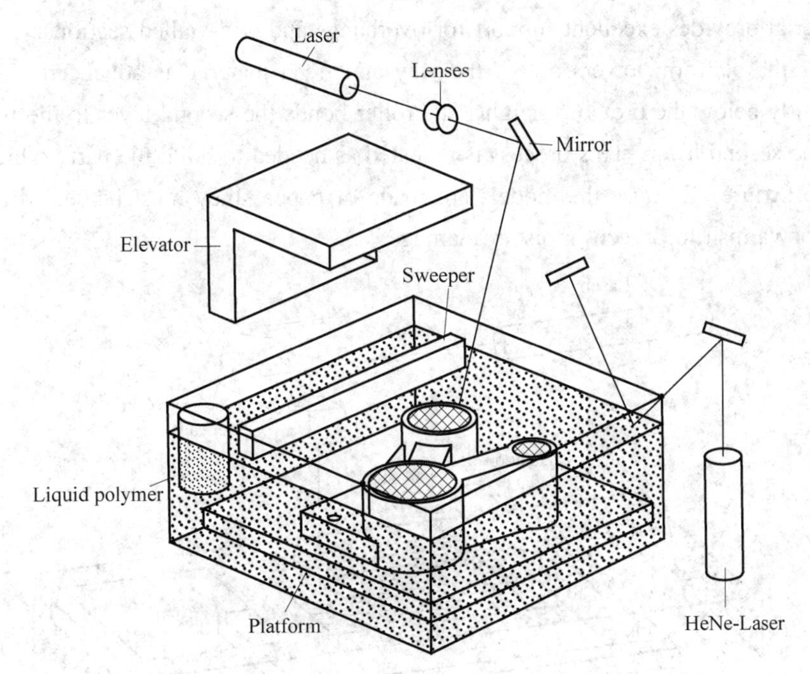

Fig. 5-12 Schematic diagram of stereolithography

Next, an elevator incrementally lowers the platform into the liquid polymer. A sweeper re-coats the solidified layer with liquid, and the laser traces the second layer atop the first. This process is repeated until the prototype is complete. Afterwards, the solid part is removed from the vat and rinsed clean of excess liquid. Supports are broken off and the model is then placed in an ultraviolet oven for complete curing.

Stereolithography Apparatus (SLA) machines have been made since 1988 by 3D Systems of Valencia, CA. To this day, 3D Systems is the industry leader, selling more RP machines than any other company. Because it was the first technique, stereolithography is regarded as a benchmark by which other technologies are judged. Early stereolithography prototypes were fairly brittle and prone to curing-induced warpage and distortion, but recent modification have largely corrected these problems.

2. Laminated Object Manufacturing

In this technique, developed by Helisys of Torrance, CA, layers of adhesive-coated sheet material are bonded together to form a prototype. The original material consists of paper laminated with heat-activated glue and rolled up on spools. As shown in Fig. 5-13, a feeder/collector mechanism advances the sheet over the build platform, where a base has been constructed from paper and double-sided foam tape. Next, a heated roller applies pressure to bond the paper to the base. A focused laser cuts the outline of the first layer into the paper and then cross-hatches the excess area (the negative space in the prototype). Cross-hatching breaks up the extra material, making it easier to remove during post-processing. During the build, the excess material provides excellent support for overhangs and thin-walled sections. After the first layer is cut, the platform lowers out of the way and fresh material is advanced. The platform rises to slightly below the previous height, the roller bonds the second layer to the first, and the laser cuts the second layer. This process is repeated as needed to build the part, which will have a wood-like texture. Because the models are made of paper, they must be sealed and finished with paint or varnish to prevent moisture damage.

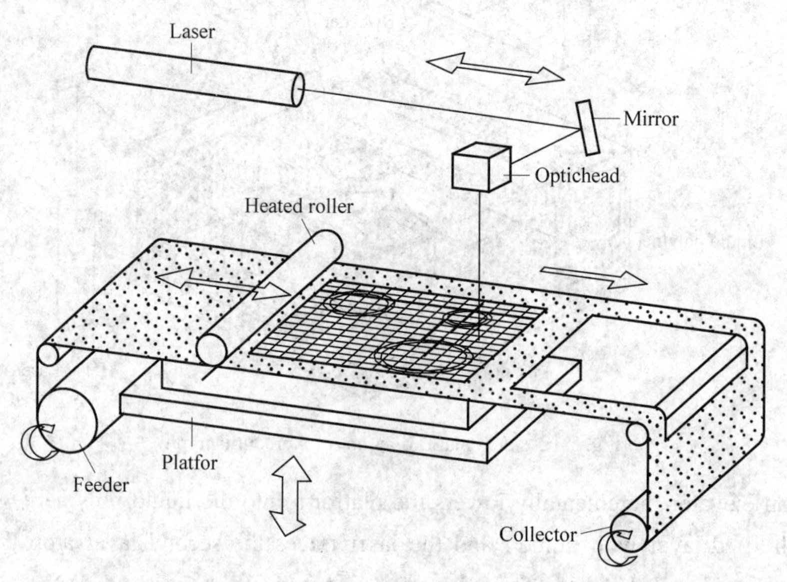

Fig. 5-13 Schematic diagram of laminated object manufacturing

3. Selective Laser Sintering

Developed by Carl Deckard for his master's thesis at the University of Texas, selective laser sintering was patented in 1989. The technique, shown in Fig. 5-14, uses a laser beam to selective fuse powdered materials, such as nylon, elastomer, and metal, into a solid object. Parts are built upon a platform which sits just below the surface in a bin of the heat-fusible powder. A laser traces the pattern of the first layer, sintering it together. The platform is lowered by the height of the next layer and powder is reapplied. This process continues until the part is complete. Excess powder in each layer helps to support the part during the build. SLS machines are produced by DTM of Austin, TX.

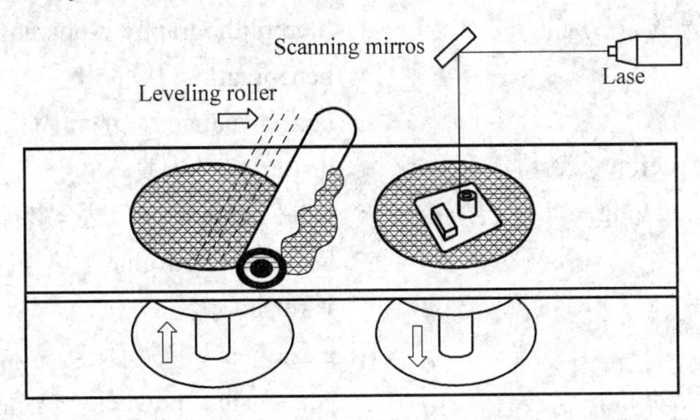

Fig. 5-14 Schematic diagram of selective laser sintering

4. Fused Deposition Modeling

In this technique, filaments of heated thermoplastic are extruded from a tip that moves in the *x-y* plane. Like a baker decorating a cake, the controlled extrusion head deposits very thin beads of material onto the build platform to form the first layer. The platform is maintained at a lower temperature, so that the thermoplastic quickly hardens. After the platform lowers, the extrusion head deposits a second layer upon the first. Supports are built along the way, fastened to the part either with a second, weaker material or with a perforated junction.

Stratasys, of Eden Prairie, MN makes a variety of FDM machines ranging from fast concept modelers to slower, high-presion machines. Material include ABS (standard and material grade), elastomer(96 durometer), polycarbonate, and investment casting wax.

New Words and Expressions

tangible 原型,有形的
physical model 物理原型
visual aids 直观教具
aerospace 航天
airfoil 机翼,螺旋桨
wind tunnel 风洞
drag force 阻力,迎面阻力
production-quality 生产
rapid tooling 快速模具
incorporating 结合
non-prototyping 非原型
software package 软件包
slice 薄片,切片
additive 叠加的
economical 经济的,节约的
wire-frame 线框
algorithm 算法
pre-existing 预存
stereolithography 立体平版印刷
planar triangle 平面三角形
vertice 顶点
manageability 可管理性
pre-processing 前处理
auxiliary 辅助的,附属的
overhang 突出物,突出部分
proprietary 专有的

autonomous 自动的
post-processing 后处理
photosensitive 感光的，光敏的
sealing 密封
sintering 烧结
Fused Deposition Modeling 熔融沉积成形
Laminated Object Modeling 实体层压成形
platform 平台
vat 大桶
epoxy 环氧树脂
acrylate resin 丙烯酸树脂
elevator 升降装置
sweeper 清洁装置
lenses 透镜
rinsed 冲洗，漂洗
Stereolithography Apparatus 立体印刷成形
benchmark 基准
curing-induced 诱导固化
distortion 失真
adhesive-coated 黏合剂涂层
heat-activated glue 热活化胶
varnish 漆
elastomer 弹性体，弹性材料
heat-fusible powder 热熔粉末
perforated 穿孔
polycarbonate 聚碳酸酯

Chapter 6

CAD/CAM/CAE

6.1 The Computer in Die Design

The term CAD is alternately used to mean computer aided design and computer aided drafting. Actually it can mean either one or both of these concepts, and the tool designer will have occasion to use it in both forms.

CAD(computer aided design) means using the computer and peripheral devices to simplify and enhance the design process. CAD(computer aided drafting) means using the computer and peripheral devices to produce the documentation and graphics for the design process. This documentation usually includes such things as preliminary drawings, working drawings, parts lists, and design calculations.

A CAD system, whether taken to mean computer aided design system or computer aided drafting system, consists of three basic components: (1) hardware, (2) software, and (3) users. The hardware components of a typical CAD system include a processor, a system display, a keyboard, a digitizer, and a plotter. The software component of a CAD system consists of the programs which allow it to perform design and drafting functions. The user is the tool designer who uses the hardware and software to simplify and enhance the design process.

The broad-based emergence of CAD on an industry-wide basis did not begin to materialize until the 1980's. However, CAD as a concept is not new. Although it has changed drastically over the years, CAD had its beginnings almost thirty years ago during the middle 1950's. Some of the first computers included graphics displays. Now a graphics display is an integral part of every CAD system.

Graphics displays represented the first real step toward bringing the worlds of tool design and the computer together. The plotters depicted in figure, represented the next step. With the advent of the digitizing tablet in the early 1960's, CAD hardware as we know it today began to take shape. The development of computer graphics software followed soon after these hardware developments.

Early CAD systems were large, cumbersome, and expensive. So expensive, in fact, that only the largest companies could afford them. During the late 1950's and early 1960's, CAD was looked on as an interesting, but impractical novelty that had only limited potential in tool design applications. However, with the introduction of the silicon chip during the 1970's, computers began to take their place in the world of tool design.

Integrated circuits on silicon chips allowed full scale computers to be packaged in small consoles no larger than television sets. These "mini-computers" had all of the characteristics of full scale computers, but they were smaller and considerably less expensive. Even smaller computers called microcomputers followed soon after.

The 1970's saw continued advances in CAD hardware and software technology. So much so that by the beginning of the 1980's, making and marketing CAD systems had become a growth industry. Also, CAD has been transformed from its status of impractical novelty to its new status as one of the most important inventions to date. By 1980, numerous CAD systems were available ranging in sizes from microcomputer systems to large minicomputer and mainframe systems.

New Words and Expressions

computer aided drafting 计算机辅助绘图
peripheral 边缘的,外表面的,周边的
documentation 文件,记录,提供文件
graphics 制图法
preliminary drawing 预制图
working drawing 工作图
parts list 零件目录表
design calculation 设计计算
hardware 硬件
software 软件
user 使用者
processor 处理器
system display 系统展示
keyboard 键盘
digitizer 数字转换器
plotter 绘图机
broad-based 无限的
industry-wide 工业领域
materialize 物化
graphics display 图形展示
digitizing tablet 数字化板
take shape 成形
cumbersome 笨重的
tool design 工具设计
silicon chip 硅片
integrated circuit 集成电路
console 仪表板,控制台
television set 电视接收机,电视机
mini-computer 小型机
microcomputer 微型(电子)计算机
mainframe system 主机系统,大型机

6.2 CAD/CAM

6.2.1 CAD

Computer-aided design/computer-aided manufacturing (CAD/CAM) refers to the integration of computers into the design and production process to improve productivity. The heart of the CAD/CAM system is the design terminal and related hardware, such as computer, printer, plotter, paper tape punch, a tape reader, and digitizer. The design is constantly monitored on the terminal until it is completed. A hard copy can be generated if necessary. A computer tape or other control medium containing the design data guides computer-controlled machine tools during the manufacturing, testing, and quality control.

The software for CAD/CAM is a collection of computer programs stored in the system to make the various hardware components perform specific tasks. Examples of software are programs developed to generate a NC tool path, to assemble a bill of materials, or to create nodes and elements on a finite element model. Some of these software packages are referred to as software modules and can be classified into four categories: (1) operating systems, (2) general-purpose programs, (3) application programs, and (4) user programs. Although there are other kinds of software, these are sufficient for an explanation of the complexities in developing a CAD/CAM system.

Operating systems are programs written for a specific computer or class of computers. For convenient and efficient operation, programs and data are available in the system's memory. The operating system is especially concerned with the input/output (I/O) devices like displays, printers, and tape punches. In most cases the operating system is supplied with the computer.

Although it may be argued that there are no general-purpose programs as such, some are more general than others. An example is a graphics program written in a high-level language like FORTRAN that allows the generation of geometric entities such as lines, circles, and parabolas and a combination of these to make designs. These designs may range from printed circuits to drill jigs and fixtures.

Application programs are developed for a special or specific purpose. The first language for specialized application was Automatically Programmed Tools (APT) in 1956. APT was developed to ease the job of NC programmers in developing input to NC machine tools, as illustrated in Fig. 6-1. Other examples of application programs, relative to CAD/CAM, are programs developed specifically for the generation of finite element mesh and flat pattern development or "unbending" of sheet metal parts. These programs are usually purchased with the system or from a software supplier.

User programs in CAD/CAM are highly specialized packages for creating specific outputs. For example, a user program may automatically design a gear after the user inputs certain parameters like the number of teeth, pitch diameter, and so on. Another program may calculate

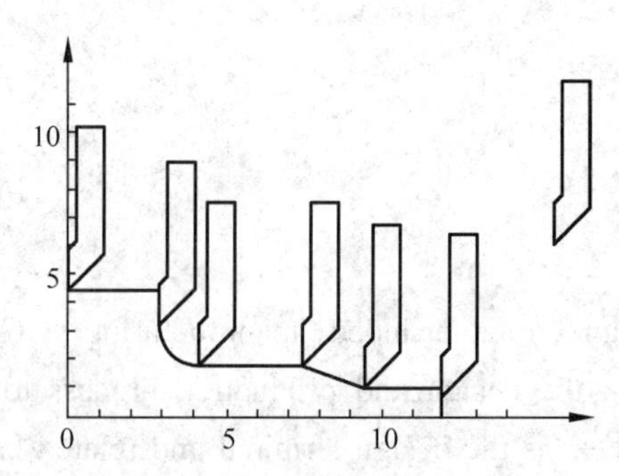

Fig. 6-1 Example of CAD where a lathe tool is called up to show a machining sequence

optimum feeds and speeds, given cutter information, material, depth of cut, and so on. These programs are often developed by the user from a software module furnished by the supplier of general-purpose software. Not all CAD/CAM software packages have these user programs, even though considerable savings can be achieved with them.

1. Computer Graphics

The computer graphics system accumulates and stores physically related data identifying the precise location, dimensions, descriptive text, and other properties of every design element. The design-related data help the user-operator perform complex engineering analysis, generate bills of materials, produce reports, and detect design inconsistencies before the part reaches manufacturing.

With computer graphics, two-dimensional drawings can be made into three-dimensional wire frames and solid models.

2. Wire Frames

The simple wire frame plot is the least expensive form of geometrically displaying a model. It is useful to verify the basic properties of a shape and continuity of the model. However, when a complex model is developed, wire frame displays become inadequate. Solid models eliminate most of the problems of the wire frame.

3. Solid Modeling

There are three basic techniques for generating solid models: constructive solid geometry (CSG), boundary representation (B-Rep), and analytical solid modeling.

In the CSG approach, various geometric patterns such as cylinders, spheres, and cones are combined by Boolean algebra to create designs.

In the B-Rep method, a profile of the part is defined and then swept, either linearly or radially, and the enclosed area represents the solid form.

4. Analytical Method

This method is similar to the B-Rep but enhances the creation of finite element model during generation of the design. Commercial packages do not use strictly one method or another. As an example, CSG packages may use B-Rep techniques to generate initial patterns, while B-Rep or analytical packages may use Boolean algebra to subtract patterns, such as cylinders or cones, from a design to create a hole in the design.

6.2.2 CAM

Computer-aided manufacture (CAM) centers around four main areas: NC, process planning, robotics, and factory management.

1. Numerical Control

The importance of NC in the CAM area is that the computer can generate a NC program directly from a geometric model or part. At present, automatic capabilities are generally limited to highly symmetric geometries and other specialized parts. However, in the near future some companies will not use drawings at all, but will be passing part information directly from design to manufacturing via a data base. As the drawings disappear, so will many of the problems, since computer models developed from a common integrated database will be used by both design and manufacturing. This can be done even though the departments may be widely separated geographically, because in essence they will be no farther apart than the terminals on their respective desks.

2. Process Planning

Process planning involves the detailed planning of the production sequence from start to finish. What is relevant to CAM is a process planning system that is able to produce process plans directly from the geometric model database with almost no human assistance.

3. Robots

Many advances are being made to integrate robots into the manufacturing system, as in on-line assembly, welding, and painting.

4. Factory Management

Factory management uses interactive factory data collection to get timely information from the factory floor. At the same time, it uses this data to calculate production priorities and dynamically determine what work needs to be done next to ensure that the master production schedule is being properly executed. The system can also be directly modified to satisfy a specific need without calling in computer programming experts.

New Words and Expressions

design terminal 终端设计　　　　printer 打印机

paper tape punch 纸带穿孔机
tape reader 磁带播放机
computer-controlled machine tools 计算机控制的工作母机
element model 单元模型
operating system 操作系统
general-purpose program 多用途程序
application program 应用程序
user program 使用程序
high-level language 高级语言
FORTRAN 公式翻译程序语言
geometric entity 几何实体
parabola 抛物线
printed circuit 印制电路
drill jig 钻孔机筛选
Automatically Programmed Tool (APT) 自动编程序系统,自动程序控制工具
element mesh 单元网格
unbending 不易弯曲的,松弛的
teeth 齿
pitch diameter 节圆直径
bill of materials 材料表
inconsistency 不一致,不合理
wire frame 线框
cone 锥体
Boolean algebra 布尔代数
process planning 制订工艺过程
factory management 工厂管理
database 资料库
in essence 本质上
on-line assembly 在线装配
painting 喷漆
computer programming expert 计算机设计专家

6.3 CAE

Computer-aided engineering simplifies the creation of the database, by allowing several applications to share the information in the database. These applications include, for example, (a) finite-element analysis of stresses, strains, deflections, and temperature distribution in structures and load-bearing members, (b) the generation, storage, and retrieval of NC data, and (c) the design of integrated circuits and other electronic devices.

With the increasing sophistication of computer hardware and software, one area which has grown rapidly is computer simulation of manufacturing processes and systems. Process simulation takes two basic forms: (a) it is a model of a specific operation intended to determine the viability of a process or to optimize or improve its performance; (b) it models multiple processes and their interactions, and it helps process planners and plant designers in the layout of machinery and facilities.

Individual processes have been modeled using various mathematical schemes. Finite-element analysis has been increasingly applied in software packages (process simulation) that are commercially available and inexpensive. Typical problems addressed are process viability (such as the formability of sheet metal in a certain die), and process optimization (for example, the material flow in forging in a given die, to identify potential defects, or mold design in casting, to eliminate hot spots, to promote uniform cooling, and to minimize defects).

Simulation of an entire manufacturing system involving multiple processes and equipment helps plant engineers to organize machinery and to identify critical machinery elements. In addition, such models can assist manufacturing engineers with scheduling and routing (by discrete event simulation). Commercially available software packages are often used for these simulations.

Polymer processing, in its most general context, involves the transformation of a solid (sometimes liquid) polymeric resin, which is in a random form (e. g. powder, pellets, beads), to a solid plastics product of specified shape, dimensions, and properties. The ultimate properties of the article are closely related to the microstructure. Therefore, the control of the process and product quality must be based on an understanding of the interactions between resin properties, equipment design, operating conditions, thermo-mechanics history, microstructure, and ultimate product properties. Mathematical modeling and computer simulation have been employed to obtain an understanding of these interactions. Such an approach has gained more importance in view of the expanding utilization of computer aided engineering (CAE) systems in conjunction with plastics processing.

6.3.1 MPI Introduction

MPI, as one of simulation software of MOLDFLOW corporation, is a set of integrated CAE simulations for plastics molding process, including MPI/Flow analysis, MPI/Co-Injection analysis, MPI/Gas analysis, MPI/Optim analysis, MPI/Microcellular analysis, MPI/Shrink analysis, MPI/Cool analysis, MPI/Warp analysis, MPI/Stress analysis. MPI provides solution in all stages of design and manufacturing, to improve productivity and enhance part quality. Key features of MPI include:

(1) Unique, patented fusion technology. MPI/Fusion allows you to analyze CAD solid models of thin-walled parts directly, resulting in a significant decrease in model preparation time. The timesaving allow you to analyze more design iterations as well as perform more in-depth analyses.

(2) Powerful workflow and productivity tools. The user-friendly environments in MPI employ visualization and project management tools that allow you to undertake extensive design analysis and optimization. After your analyses are complete, you can produce detailed, web-ready design reports quickly and easily.

(3) Proven solutions for all types of applications. Moldflow's analysis products can simulate plastics flow and packing, mold cooling, and part shrinkage and warpage for thermoplastic injection molding, gas-assisted injection molding, co-injection molding and injection- compression molding processes. Additional modules simulate reactive molding processes including thermoset and rubber injection molding, reaction injection molding (RIM), structural reaction injection molding (SRIM), resin transfer molding, microchip encapsulation and underfill (flip-chip) encapsulation.

(4) The world's best 3D solution. Using a proven solution technique based on a solid tetrahedral finite element volume mesh, MPI/3D allows you to perform true 3-dimensional flow

simulations on parts that tend to be very thick and solid in nature as well as those that have extreme changes from thin to thick.

(5) The widest range of supported geometry types. Moldflow's MPI technology can be used on all CAD model geometry types, including traditional midplane models, wire frame and surface models, thin-walled solids and thick or difficult-to-midplane solids. Regardless of your design geometry, you can accomplish simulation tasks in an easy-to-use, consistent, integrated environment that works with your model.

(6) Unsurpassed network computing options. Moldflow Plastics Insight (MPI) has been developed with network computing environments in mind. For example, users can run MPI/Synergy, the pre- and post-processor, on user-friendly Windows PC while running the analysis solvers on powerful UNIX workstations. Users can also take advantage of a distributed computing environment from within MPI and assign analyses to run on whatever network computers are available at any given time.

6.3.2 MPI Modules

1. MPI/Flow

MPI/Flow simulates the filling and packing phases of the injection molding process to predict the flow behavior of thermoplastic melts so you can ensure parts of acceptable quality can be manufactured efficiently.

Using MPI/Flow, you can

- predict and visualize the flowfront progression to see how the mold fills;
- determine injection pressure and clamp force requirements;
- optimize part wall thickness to achieve uniform filling;
- minimize cycle time, and reduce part cost;
- predict weld line locations and either move, minimize, or eliminate them, identify potential air traps and determine locations for proper mold venting;
- optimize process conditions such as injection time, injection velocity profile, melt temperature, packing pressure, packing time, and cycle time;
- determine areas of high volumetric shrinkage that could cause part warpage problems;
- determine gate freeze time.

2. MPI/Cool

MPI/Cool provides tools for modeling mold cooling circuits, inserts, and bases around a part and analyzing the efficiency of the mold's cooling system.

MPI/Cool simulations allow users to:

- optimize part and mold designs to achieve uniform cooling with the minimum cycle time;
- view the temperature difference between the core and cavity mold surfaces;
- minimize unbalanced cooling and residual stress to reduce or eliminate part warpage;

- predict temperature for all surfaces within the mold: part, runners, cooling channels, inserts;
- predict the required cooling time for the part and cold runner to determine overall cycle time.

3. MPI/Warp

MPI/Warp provides users with an understanding of the causes of shrinkage and warpage in injection molded plastic parts and predicts where deformations will occur.

MPI/Warp allows you to:

- evaluate final part shape before machining the mold;
- evaluate both single cavity and multi-cavity molds, scale shrinkage and warpage results for better visualization of deformation;
- query any two points to determine any dimensional change between the two;
- constrain the part on a plane for better measurement of deflection;
- separate total displacement into *X*-, *Y*-, and *Z*-axis displacements to show only the deflection in each direction;
- show shrinkage and warpage as a visible displacement plot or as a color contour or shaded plot;
- export warp geometry in the STL format to use as a reference when sizing the mold;
- export warp mesh model for an iterative warpage analysis.

4. MPI/Fiber

MPI/Fiber predicts the fiber orientation due to flow in fiber-filled plastics and the resultant mechanical strength of the plastic/fiber composite.

It is important to understand and control the orientation of fibers within fiber filled plastics to reduce shrinkage variations across the molded part to minimize or eliminate part warpage.

MPI/Fiber allows you to:

- predict fiber orientation and thermo-mechanical property distributions in the molded part;
- predict elastic modulus and average modulus in the flow and transverse-flow directions;
- predict linear thermal expansion coefficient (LTEC) and average LTEC;
- calculate Poisson's Ratio, a measure of the transverse contraction of a part compared to its length when exposed to tensile stress;
- optimize filling pattern and fiber orientation to reduce shrinkage variations and part warpage;
- increase part strength by inducing fiber orientation along load bearing part surfaces.

5. MPI/Optim

MPI/Optim allows you to perform an injection molding machine-specific analysis which takes into account the actual machine response time, maximum injection velocity, and the number of steps that can be programmed for velocity and pressure profiles on the machine controller. The analysis aim is to achieve uniform flow front velocity and temperature profiles through the injection molding machine nozzle, the mold feed system, and the part cavities.

MPI/Optim allows you to:

- automatically determine the optimum processing conditions, including stroke length, injection velocity profile, velocity-to-pressure switch-over, and pressure profile, required to produce a part of acceptable quality given a mold, machine, and material combination;
- implement the optimum setup conditions directly on the molding machine;
- take into consideration key quality criteria for the part, including control of short shot, flash, warpage, the desired dimensional tolerance, flow front velocity, constant flow front temperature, and frozen layer thickness.

6. MPI/Stress

MPI/Stress is a structural analysis product for optimizing the structural integrity of plastic injection molded parts. It considers the effects of plastic flow during injection molding and the resultant mechanical properties on the component's structural integrity, ensuring it is fit- for-purpose and will not fail in use, well before mold trials or production, when the cost of change is high.

MPI/Stress allows you to:

- predict stresses and product deflections that result from external loads and temperatures;
- account for the effects of processing on mechanical properties of the part and the orthotropic (direction dependent) properties of injection molded components;
- evaluate whether a structural part, previously made from metal or other materials, can be successfully made from plastic;
- enable iterative optimization of part design to ensure the molding will meet final product strength and stiffness specifications;
- eliminate the need to over-engineer parts, using unnecessary costly engineering materials and thicker wall sections to achieve structural requirements.

7. MPI/Shrink

MPI/Shrink predicts polymer shrinkage based on the effects of processing and grade-specific material data and offers a true prediction of linear shrinkage independent of warpage analysis. Because plastic parts shrink as they cool, it is essential to accurately account for this shrinkage in the design of the mold so that critical product tolerances can be met.

MPI/Shrink allows you to:

- provide precise, optimum shrinkage values and predict shrinkage variations across the mold, so that mold design can be refined to compensate for these variations;
- enable control of molding conditions, gate location, and material grade selection to

ensure specified part dimensions will be achieved;
- understand the effects of processing on shrinkage;
- replace the traditional "best-guess" approach for determining shrinkage values;
- eliminate the need to cut the mold under size and re-machine to finished size after mold trials;
- eliminate the need to push molding conditions to the limit to achieve the required dimensions;
- ensure the mold produces parts that are within critical tolerances, thereby reducing reject rates;
- eliminate the need to adjust molding parameters to give the required dimensions, which often results in molding far from optimum conditions, resulting in long cycle times and surface defects;
- decrease the need for prototype tooling;
- evaluate the performance of different materials.

8. MPI/Gas

MPI/Gas simulates the gas-assisted injection molding process, where gas, usually inert nitrogen, is injected into the polymer melt. The gas drives the polymer through the mold cavity to complete mold filling and create a network of hollow channels throughout the component. Combine MPI/Gas with MPI/Cool, MPI/Fiber, and MPI/Warp to help determine where to put polymer and gas entrances, how much plastic to inject prior to gas injection, where to place gas channels, and how large to size them.

MPI/Gas allows you to:
- evaluate the filling pattern with the influence of gas injection to aid in part design, gate placement, and process setup;
- link to MPI/Cool to evaluate mold cooling with the influence of gas injection to optimize mold cooling design and minimize cycle time;
- link to MPI/Warp to predict part shrinkage and warpage with the influence of gas penetration to determine final part quality;
- link to MPI/Stress to apply in-service loading to determine part performance with gas channels;
- properly size gas channels for optimal filling and gas penetration;
- determine the best gas channel layout to control gas penetration, inject gas at any location or in multiple locations within the part or runner system;
- inject gas through multiple gas pins simultaneously or at different times during the process;
- detect areas of poor gas penetration or other problems, determine the proper shot size to avoid gas "blowout";
- optimize injection speed profile for the plastic injection stage;

- determine injection pressure and clamp force requirements for proper molding machine selection;
- incorporate delay time prior to injecting gas allowing thin areas to solidify;
- automatically determine gas pressure required to avoid short shots, melt-front hesitation, or burning;
- determine final part weight after gas injection to help maximize material savings and minimize weight;
- estimate the final wall thickness after gas penetration;
- accurately identify weld (knit) and meld lines based on part design and gate placement
- accurately identify air traps for proper mold venting.

9. MPI/Co-Injection

MPI/Co-Injection is an ideal molding process for using recycled materials or achieving specialized cosmetic and structural objectives. MPI/Co-Injection provides an invaluable tool for simulating the sequential co-injection process, where a skin material is injected first, followed by the injection of a different core material.

MPI/Co-Injection allows you to:

- evaluate the flow front pattern of two co-injected materials to aid in part design and gate placement;
- predict the extent of penetration of the core material and whether it will break through the skin material;
- determine injection pressure and clamp force requirements for proper molding machine selection;
- balance and minimize runner systems to achieve uniform cavity filling with reduced scrap or regrind material;
- determine the best transition point for switching from skin-material injection to core-material injection;
- place gate locations to minimize injection pressure and clamp force;
- simulate different inlet melt temperatures for skin and core materials;
- automatically incorporate the recommended ram-speed profile from MPI/Flow to reduce overshearing of the plastic during filling;
- accurately identify weld and meld lines based on part design and gate placement.

6.3.3 CAE Example of MPI

This example illustrates various aspects of the power CAE software for the plastics industry, and how such software can be strategically applied. It illustrates the role of computer simulation at each iterative step in process.

1. Flow analysis

The basic procedures for doing a Filling or Flow analysis are similar even with the wide variety of objectives. A typical analysis may include the following steps:

(1) Import the CAD model.

This is simply a file read function. It is just like opening a document in a word processor. In the case of MPI, reading in a format like STL or IGES, the geometry comes in but it can't be used for analysis until it is meshed. In the case of mesh formats like Patran, the mesh is cleaned up and the model can be ready to run when the injection location, material and processing conditions are set.

(2) Translate the CAD model (generate mesh).

With file formats such as STL and IGES, an initial finite element mesh must be created. After the initial mesh is created and inspected, a decision could be made that the mesh density is too fine or too course. The original mesh is then deleted and another global edge length is defined.

(3) Refine the mesh density.

The mesh may be fine in most areas of the part but not in a specific area of the part. In some smaller model detail the mesh could be too coarse. The local mesh can be changed using the Mesh tools to refine the local mesh density.

(4) Clean up the model.

After the mesh density has been determined, the mesh may need to be cleaned up. Normally some cleanup is necessary if the mesh was not translated in from another system. There is a mesh statistics report used to check the quality of the mesh, a series of diagnostic displays to show problems, and mesh tools to repair any mesh problems.

(5) Select the material.

For any analysis, the particular material to be used must be found in the material database. There are several techniques to find a specific grade of material, assess the quality of the material data, find a substitute material, and compare materials within the Select Material dialog.

(6) Select the gate location.

A gate location on a part may be fixed, there may be two or three choices, or the optimum gate(s) locations may need to be found as a major part of the analysis process. In any case, the injection location needs to be selected initially. There may be many iterations involving running a fill analysis, or even a flow analysis to decide on the final gate location(s).

(7) Select the molding machine.

There may be cases where the specific molding machine information is required. The default molding machine has a pressure limit of 100 MPa. This is a good design limit. The clamp tonnage limit is set very high, at 7,000 tonnes. However, if the actual injection molding machine has a higher pressure limit than the standard machine, or if the clamp tonnage could be an issue, a specific molding machine should be selected.

(8) Determine molding conditions.

The molding conditions to be used may be mandated at least as a starting point. The molding conditions to be used can be determined by a molding window analysis and modified slightly in a filling analysis. A molding window analysis can be used as a quick initial analysis to compare materials and gate locations in addition to optimization of the molding conditions. It can save a significant amount of time if good processing conditions are chosen before a fill analysis is done.

(9) Set analysis parameters.

Analysis parameters in addition to the 'molding conditions' include the velocity/pressure switch-over, the packing profile, the cooling time, solver settings, process controller settings, the mold material and others.

(10) Run the analysis.

This refers to normally a fill analysis or a flow analysis. A fill analysis stops when the part volume is just filled to 100%. The flow analysis is a fill analysis but continues through the packing and even cooling phases of the molding cycle.

(11) Solve the filling problem.

This is not any one analysis; this is an iterative process. Once the first fill analysis is done, the results are reviewed, and a problem is identified and fixed. This may require many iterations of the previous steps.

(12) Runner balancing.

Once the filling is optimized, the runner system can be analyzed, sized, and balanced. This includes sizing the gate and sprue. Depending on the objectives of the project, this step may not be done. It could end once the filling is done. However, to do a complete analysis at least a proposed runner system should be analyzed.

(13) Determine the packing profile.

Once the runner system is sized, the packing can be investigated. Although a flow analysis can be done without a gate or runner, it is not recommended if you are interested in how the part is going to be packed out. The packing of the part is significantly affected by the gate and runner freeze time. Without a runner and gate, the packing analysis will be less accurate.

Following figures (Fig. 6-2 and Fig. 6-3) display the results of flow analysis:

2. Cool Analysis

MPI cooling analyses are heat transfer analysis products designed to analyze flow of heat in plastic injection molds. The principle outcomes of an analysis are temperatures in the plastic filled cavity and throughout the mold, cooling time, and cooling network flow parameters such as pressure or flow rate requirements. Network analysis also provides information on pumping requirements for a given circuit and coolant combination.

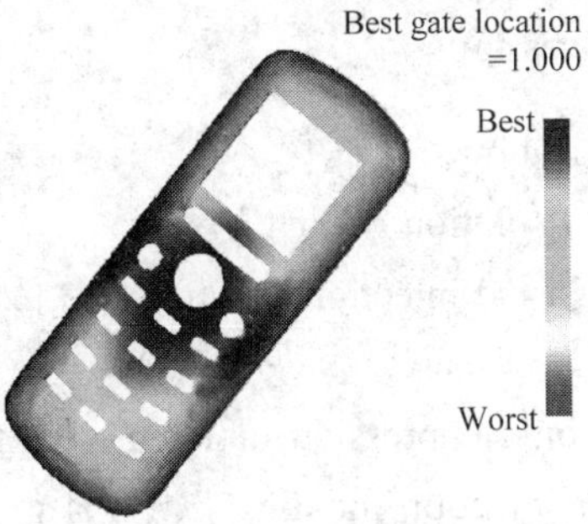

Fig. 6-2 Best gate location

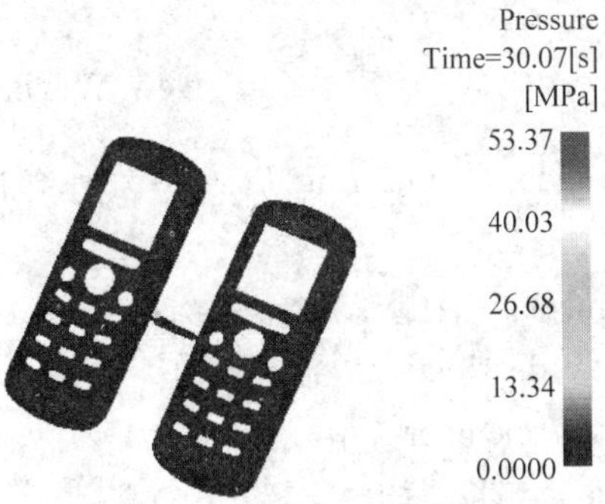

Fig. 6-3 Pressure profile at end of injection

A feature of the MPI cooling analyses is that they interface to the MPI/Flow analyses. This enables the flow analysis to recognize the effects of local hot and cold spots arising from a given cooling situation. MPI/Warp, in turn, considers the cooling effects carried onto the flow analysis in order to compute the impact of differential temperature distributions on part warpage (only available for Midplane models).

Following figures (Fig. 6-4, Fig. 6-5 and Fig. 6-6) display the results of flow analysis:

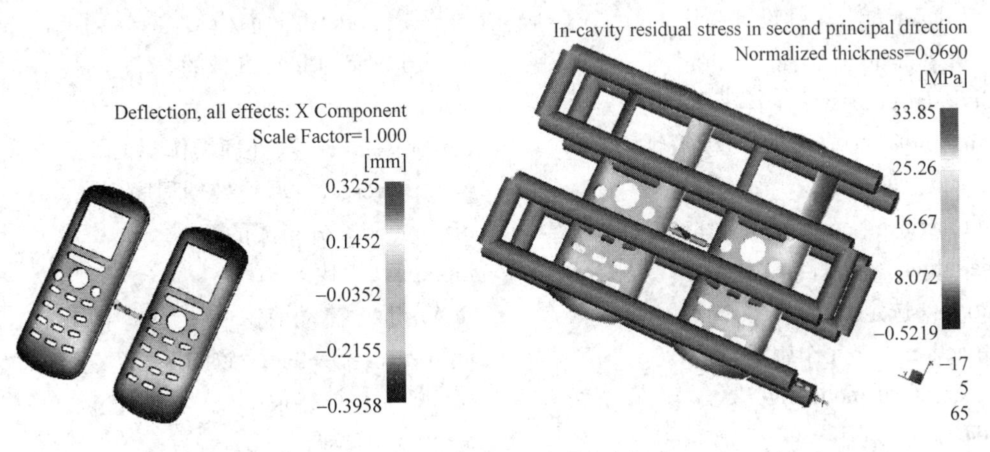

Fig. 6-4 Deflection in X direction

Fig. 6-5 In-cavity residual stress

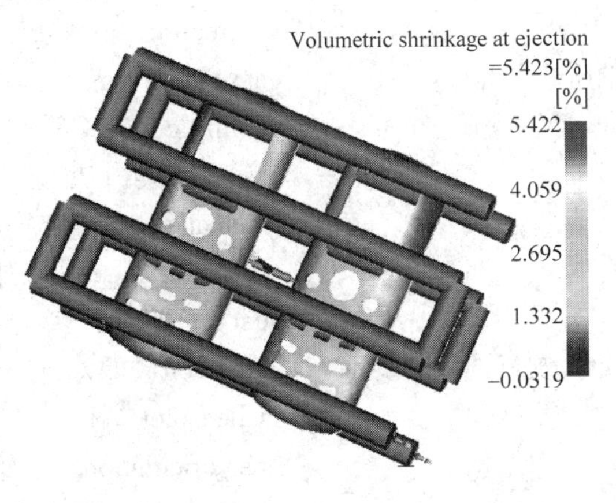

Fig. 6-6 Volumetric shrinkage at ejection

New Words and Expressions

computer-aided engineering（CAE） 计算机辅助工程
finite-element 有限元
temperature distribution 温度分布
load-bearing member 承载元
generation 产生，发生
retrieval 恢复，修补
electronic device 电子设备
sophistication 混合
computer simulation 计算机模拟
process simulation 过程模拟
viability 生存能力，发育能力
multiple process 复合过程
process planner 过程计划者
plant designer 车间设计者
layout 布局图，版面设计
individual process 个体处理
formability 可成形性，可模锻性
hot spot 加热部位
manufacturing engineer 制造业工程师
scheduling 编制目录
routing 选定路线
interaction 交互作用
mathematical modeling 数学建模
utilization 利用
microcellular 微分网格
key feature 主要特征
patented fusion 独特的融合技术
solid model 实体模型
thin-walled part 薄壁制件
iteration 反复
in-depth 深入的
workflow 工作流
user-friendly environment 友好的用户界面
visualization 可视化
project 计划，工程
undertake 承担，采取
optimization 最佳化
proven solution 证明方法
gas-assisted injection molding 气体辅助注射成形
injection-compression molding 挤塑成形
rubber injection molding 橡胶注射成形
structural reaction injection molding（SRIM） 结构反应注射成形
microchip encapsulation 微芯片封装
underfill（flip-chip）encapsulation 封装
tetrahedral 四面体的
volume mesh 体积网格
geometry type 几何类型
midplane model 中型面模型
surface model 曲面模型
unsurpassed 非常卓越的
synergy 协同，配合
pre-processor 预处理操作
post-processor 后置处理程序
distributed 分布式的
module 模块
flowfront 流动前锋
uniform 统一的，一致的
weld line 熔接痕
air trap 气泡
injection velocity profile 注射速度分布图
volumetric shrinkage 体积收缩量
gate freeze time 浇口凝固时间
cooling circuit 冷却管道
base 模架
residual stress 残余应力
cooling channel 冷却管道
query 询问
deflection 偏差，偏转
shaded plot 阴影图
fiber orientation 纤维取向
transverse-flow 横流

linear thermal expansion coefficient (LTEC) 线性热膨胀系数
transverse contraction 横向收缩
tensile stress 张应力
machine-specific 指定的设备
take into account 考虑，重视
stroke length 行程长度
velocity-to-pressure switch-over 速度－压力转换点
pressure profile 压力分布图
short shot 短射
frozen layer 凝固层
orthotropic 正交各向异性的
over-engineer 过加工
grade-specific 指定的等级
material data 材料参数
compensate 补偿
best-guess 最佳猜想
re-machine 再加工
mold trial 试模
reject rate 废品率
prototype 原型
in-service 在用的
blowout 喷出
delay time 延迟时间，滞后时间
melt-front 熔体前沿
hesitation 滞留
burning 燃烧，氧化
cosmetic 化妆品
sequential 连续的，有顺序的
skin material 表面材料
regrind material 再研磨材料
transition point 过渡点
ram-speed profile 滑块速度分布图
overshearing 过剪切
mesh density 网格密度
global edge length 总边长
clean up 清除
diagnostic 诊断
material database 材料数据库
substitute material 代用材料
mandate 命令
solver setting 解决方案设置
process controller setting 过程控制设置

Glossary

（模具设计与制造专业词汇总表）

abrasion　磨损
abrasive wheel　砂轮,磨轮
absolute programming　绝对值编程
acceleration　加速
accelerator　加速器
accessories　配件
accommodate　能容纳,可搭载
accuracy　精度
acrylate resin　丙烯酸树脂
acting force　作用力
adaptive control　自适应控制
additive　叠加的
adhesive-coated　黏合剂涂层
aerospace　航天
aesthetic　工艺的
aggregate　成套设备,机组
air carbon-arc torch　空气电弧焊炬
air trap　气泡
airfoil　机翼,螺旋桨
algorithm　演算法
aligning pin　定位销
alkyd　醇酸树脂
alloy　合金
alphanumeric data　文字数字数据
aluminum　铝
ammeter　电流计,安培表
analogous　类似于
analogy　比拟,类比
anchor　固定
angle pin　斜导柱
angular　角度,用角度测量
anneal　退火
anode　阳极
anodically polarize　阳极极化
anvil　基准面
apparatus　仪器
apparent viscosity　表观黏度
appliance　设备,工具
application program　应用程序
aqueous sodium chloride　氯化钠溶液
arbitrary　任意,随意
article　项目,产品
asbestos　石棉
asterisks　星号,加星号的
attachment　附件,配件,装配装置
automated machine tool　自动化机床
automatic material handing system　自动送料系统
automatic transfer device　自动转换装置
automatically machining　自动加工
Automatically Programmed Tool（APT）　自动编程序系统,自动程序控制工具
automation　自动操作
automatization　自动化
automobile　汽车
automobile parts　汽车零部件
automotive component　自动推进部件
autonomous　自动的
auxiliary　辅助的,附属的
auxiliary ram　辅助活塞
axes　轴线
axis-symmetrical　轴对称的
backbone　中枢,支柱
backflow　逆流
backlash　后座
backup　阻塞,后援
ball bearing　球轴承
ball screw　滚珠丝杠
bank sand　岸砂(黏土少于5%的天然砂,铸造用砂)
barrel　料筒
base　模架
base support　底板,底座
batch　一批
bathtub　浴盆
bead　微珠,空心颗粒
bearing cage　轴承座

bearing clearance　轴承间隙
bed　机床，床身
bedways　床身导轨
benchmark　基准
bending　弯曲
best-guess　最佳猜想
bidirectional　双向的
bill of materials　材料表
binder　黏结剂，曲面压碎圈，双动压力机外滑块
Bingham　宾汉姆
blade　叶片，刀刃
blank　毛坯，坯料
blanking　落料，冲裁
blasting　爆炸，爆破
blend　混合
blind riser　暗冒口
blow molding　吹塑成形
blowout　喷出
blueprint　蓝图，设计图
bolt　螺栓
bond　黏合
Boolean algebra　布尔代数
boring　镗孔，镗削
brass　黄铜
breakage　破损
breaker plate　机头体
brittleness　脆性
brittle-type　脆性断裂
broad-based　无限的
bronze　青铜
buckle　纵弯，皱纹
building block　标准部件，结构单元
bulging　胀形，起凸
burning　燃烧，氧化
burr　毛刺，飞边
burr-free　无毛刺
bushing　衬套，导套
calculation algorithm　计算法则
calendering　压延，压制成形
camera　照相机
capacitance-type　电容式
carbon atoms　碳原子
carburetor　汽化器
carriage　滑动架
cascade waterway　串联水路
casting　铸造，浇注
casting cycle　压铸周期，压铸循环
cast-iron　铸铁
cast-iron casting　铸造件
category　种类，类别
cathode　阴极
cavity plate　型腔固定板
cellulose　纤维素
cellulose acetate　醋酸纤维素
centerline　中心线
center-pin-point gating　点浇口浇注系统
central axis　中心轴
centrifugal casting　离心铸造法
changeover　转换
channel　流道
chaplet　型芯撑
charge　负荷，装料
cheek　耐火侧墙
chemical reaction　化学反应
chip-removal　排屑
chop　砍，剁碎
circular angle　圆角
circumference　圆周，周边
circumferential　圆周的
circumventing　规避，绕行
clamp　压紧，夹紧
clamping　合模
clamping pad　夹紧垫片
clay　黏土，泥土
clean up　清除
clearance　间隙
clockwise　顺时针方向的，右旋的
close tolerance　小公差，紧公差
CNC（computer numerical control）　数控机床
coated sheet metal　喷涂的板材
coating　涂层，涂料
cobalt　钴
coefficient　系数
coerce　强制，迫使
coiled　盘绕
co-injection molding　共注射成形

cold shuts 冷寒,冷疤
cold-box mold 低温铸模
cold-chamber 冷压室
cold-setting process 冷塑(固)化过程
collapsibility 崩溃性,退让性
column 立柱
compact 压紧,压实
comparison circuit 比较电路
compensate 补偿
complementary 互补的
compound die 复合模
compression 压缩
compression molding 压缩成形
compressor 压缩机
computer aided drafting 计算机辅助绘图
computer integrated manufacturing 计算机一体化制造
computer numerically controlled 计算机数字控制的
computer programming expert 计算机设计专家
computer simulation 计算机模拟
computer-aided design(CAD) 计算机辅助设计
computer-aided engineering (CAE) 计算机辅助工程
computer-aided manufacturing (CAM) 计算机辅助制造
computer-controlled machine tools 计算机控制的工作母机
concave 凹形,凹面
conductive fluid 导电流体
conductivity 电导率
cone 锥体
console 仪表板,控制台
consolidation 压实,压密
constant rate 恒定速度
constant stress 恒定应力
container 容器
continuous mass 持续大量
contour 轮廓,外形
control box 操纵盒
convex 凸形
conveyor 传送带,输送机
coolant 冷却剂,冷却液
cooling 冷却
cooling channel 冷却管道
cooling circuit 冷却管道
cooling phase 冷却阶段
coordinate system 坐标系统
coordinate worktable 坐标工作台
cope 上型箱
copper 铜
copper-based alloy 铜基合金
core 芯子,型芯
core blower 芯型吹砂机,吹芯机
core locking wedge 型芯锁楔
core plate 中心板
core print 型芯座
corrosion 侵蚀
corrosion resistance 抗腐蚀性
corrosive 腐蚀的,腐蚀性的
corrugation 波纹,皱折,起皱
cosmetic 化妆品
cotton cloth 棉布
cotton linter 棉绒
counteract 抵消,减少
counterclockwise 逆时针方向的,左旋的
counterpart 对应的
cover die 定模
crack 裂缝,破裂
crankshaft 曲轴,曲柄轴
crater 有坑,形成坑
cross-link 交叉结合
cross-linked molecule 交联的分子
cross-section 横截面
crosswise 成十字状的,交叉的
crystallite 结晶性(度)
cull 精选
cumbersome 笨重的
curing 硫化,固化
curing-induced 诱导固化
curling 卷边,卷曲
current density 电流密度
curvature 弯曲,曲率
curve 曲线
curved-surface 曲面
cushion ring 垫片

custom-engineered equipment 工程设备
cutter compensation 刀具补偿
cutting edge 剪刃,刀刃,刃口
cutting process 分离工序
cycle time 周期
cylinder 圆筒,圆柱体
database 资料库
deactivate 撤销,停用
debris 碎片,残骸
deburring 去毛刺
deceleration 减速
deep drawing 深拉延
defect 过失,缺陷
deflection 偏差,偏转
delay time 延迟时间,滞后时间
describing 沿……运行,形成……形状
design calculation 设计计算
design terminal 终端设计
designation 命名,指示
detrimental effect 不利影响
diagnostic 诊断
diamond particle 金刚石颗粒
dictate 控制,支配
die 模具,砧子,凹模
die block 模板
die insert 压模嵌件
die insert waterway 压模嵌件水路
die life 模具寿命
die shoe 模座
die-casting 压力铸造,压铸
die-casting mold 压铸模
dieforming 模具成形
dielectric 绝缘体,非传导性的
dielectric fluid 介电液体,电介质
digitizer 数字转换器
digitizing tablet 数字化板
dilatant 膨胀剂
dimensional accuracy 尺寸精度
dimensional instability 尺寸不稳定性
direct current 直流电
discoloration 变色,污点
disintegration 解体
diskette 磁盘
distortion 扭曲,变形
distortion 失真
distributed 分布式的
documentation 文件,记录,提供文件
dowel pin 定位销
drag 阻力,制动,牵制
drag force 阻力,迎面阻力
drawback 缺点,障碍
drawback 缺陷
drawing of the part 零件图
drill cycle 钻孔
drill jig 钻孔机筛选
drive motor 驱动机构,传动装置
dry binderless sand 干燥的无黏结的型砂
ductile 易延展的,柔软的
ductility 延展性
dull 钝的,不锋利的
dwell 暂停
dwell time 保压时间
dye 染,染色
economic justification 经济因素
economical 经济的,节约的
ejecting force 顶出力
ejecting temperature 脱模温度,取出温度
ejector 推杆,顶出器
ejector housing 注塑模顶杆空间
ejector mechanism 顶出机构
ejector pin 推杆,顶杆
ejector return pin 顶杆回程销
elastic deformation 弹性变形
elastic modulus 弹性模量
elastic solid 弹性体
elasticity 弹性,弹力
elastomer 弹性体,弹性材料
Electric Discharge Machining (EDM) 电火花加工
electrochemical 电化学的
electrochemical dissolution 电化学溶解
electrochemical grinding 电解磨削
electrochemical machining 电解加工
electrode 电极
electrode wire 电极丝
electrolysis 电解作用

electrolyte 电解质,电解液
electrolytic cell 电解槽
electron 电子
electronic device 电子设备
electronic microscope 电子显微镜
electronic transfer 电子传递
electroplating 电镀
element mesh 单元网格
element model 单元模型
elevator 升降装置
embedded 嵌入的,内嵌的
emboss 凸起
empirical relationship 经验关系
encoder 编码器
encompass 包围,环绕
engine block 发动机组(本体)
engineering science 工程科学
entangle 卷入,纠缠在一起
epoxy 环氧树脂
epoxy resin 环氧树脂
equilibrium 平衡,均衡
equipment configuration 设备装置
ethylene 乙烯
exotic 特殊
explosion-proof blower 防爆风机
extrudate 挤出物
extruder 挤压机
extrusion 挤压
fabrication 制作,构成,伪造物,装配工
factory management 工厂管理
Faraday's law 法拉第定律
fastening 紧固
fastening plate 固定板,连接板
feed drive mechanism 进给驱动机构
feed hopper 进料斗
feed mechanism 进给机构,进料机构
feed override 进给倍率
feedback signal 反馈信号
feedback transducer 反馈传感器
ferrous 有色金属
fiber orientation 纤维取向
filament 细丝
file cabinet 文件柜
filler 装填物,漏斗,衬垫,焊补料层
film 薄膜
filter 过滤器
finishing 精加工,抛光
finite-element 有限元
fixed automation 固定自动机
fixed core 固定型芯
fixed-stripper 固定式卸料板,刚性卸料板
fixture 夹具,固定装置
fixtured 固定的
flange 凸缘
flanging 翻口,翻边,弯边
flash 飞边
flask 型(砂)箱
flexibility 弹性,机动性,挠性
flexible automation 柔性自动化
flexible manufacturing system (FMS) 柔性制造系统
floating plate (平板机的)中间热板
flood coolant 水冷或液体冷却
flowfront 流动前锋
fluid flow 流体流量
flushing 清洗
foamed part 泡沫部分
foil 铝箔
force 力
forging 锻造
formability 可成形性,可模锻性
forming process 成形过程,成形工艺
formulations 配方
FORTRAN 公式翻译程序语言
foundry 铸造,翻砂,铸造厂
fraction 小部分
fracture 断裂,断裂面
fracture zone 断裂区
friction 摩擦,摩擦力
frozen layer 凝固层
frying pan 煎锅,长柄平锅
full-form shaping 整体造型
functional 功能的
furnace 熔炉
Fused Deposition Modeling 熔融沉积成形
gantry-type tool changer 龙门刀具变换器

gas-assisted 气体辅助的
gas-assisted injection molding 气体辅助注射成形
gaseous 气体的,气态的
gasified 气化
gate 浇口
gate freeze time 浇口凝固时间
gauge 定位装置
gauging operation 测量
gear 齿轮,传动装置
gear reducer 齿轮减速装置
general-purpose program 多用途程序
generation 产生,发生
geometric element 几何元素
geometric entity 几何实体
geometry 几何学,几何形
geometry type 几何类型
glass fiber 玻璃纤维
global edge length 总边长
gooseneck 鹅颈管,流道
grade-specific 指定的等级
grain direction 晶粒方向
grain orientation 晶粒取向
grain size 晶粒度
granule 颗粒,细粒
graphic 图解的,坐标式的
graphical assistance 图形帮助
graphics 制图法
graphics display 图形展示
graphite 石墨
green-sand mold 湿(砂)型
grinding 磨,碾
grinding wheel 砂轮
grit 粗砂,研磨
groove 槽,模膛
group technology 成组工艺,成组技术
guide bushing 导套
guide pillar 导柱
guide pin 导针,导柱
guide plate 导板
guide pulleys 引导滑轮
guide roller 导轮
gun barrel 炮筒,枪筒
hardware 硬件
hardwire 硬件,固定线路
heat transfer 热传导,传热,导热
heat-activated glue 热活化胶
heater band 电热丝
heat-fusible powder 热熔粉末
heat-sensitive 热敏的
heat-treated 对……进行热处理
heavy-duty machining 重型机械
hemming 折边,卷边
hesitation 滞留
high-frequency 高频
high-level language 高级语言
high-molecular-weight 高分子量
holding 支持,把握
holding chamber 蓄料室
hole drilling 钻孔
hollow 孔,空穴
Hookean 胡克
hopper 料斗
hot probe 热探测器
hot spot 加热部位
hot-chamber 热压室
hot-manifold mold 热流道模具
hot-runner mold 热流道模具
house-hold 家用的
housing 机架
hypodermic needle 注射针
hypotenuse (直角三角形的)斜边,弦
impeller 叶轮
imperfection 缺点,瑕疵
imprint 留下烙印
impulse power 脉冲电源
in essence 本质上
inconel 铬镍铁合金
inconsistency 不一致,不合理
incorporating 结合
incremental programming 增量值编程
indentation 压入,压痕
in-depth 深入的
individual process 个体处理
industrial robot 工业机器人
industry-wide 工业领域
inertia 惯性,惯量

inertial force 惯性力
inflate 使充气,膨胀
ingredient 成分,因素
initial investment 最初投资
injection 注射
injection chamber 注射室
injection filling 注射填充
injection mold 注射模
injection molding 注射成形,注射模塑
injection ram 注射活塞
injection velocity profile 注射速度分布图
injection-compression molding 挤塑成形
inlet 入口
inorganic 无机的
in-process 进程中
input media 输入设备
in-service 在用的
insoluble 不能溶解的
instantaneous 瞬时的
instate 任命
insulated hot-runner 绝热保温流道
insulating abrasive 绝缘磨料
insulation 绝缘
insulator 绝缘体,绝缘器
integrated circuit 集成电路
integration 集成,一体化
intensity 强度
interaction 交互作用
interchange 互换,相互交换
interchangeably 可交换地,可替换地
inter-electrode 两极之间的
interior 内部的
interlock 互锁
intermittent 间歇的,断续的
intermolecular slippage 分子间滑动
interpolation 插入,插补
interpolation operation 插补运算
inverse-time 时间倒数
ion 离子
ionization 电离
ionize 电离
iteration 反复
jacket 罩,套,壳,盖
jaw 爪,夹持零件
just-in-time delivery 即时运输
key feature 主要特征
keyboard 键盘
kink 扭结,弯曲
knee 弯头,拐弯处
knob 旋钮,球形捏手
knockout pin 顶出杆
labor-intensive 劳动密集的
ladling 浇注
lake sand 湖沙
laminated 分层的
Laminated Object Modeling 实体层压成形
lancing 切缝,切口
lateral 横向的,侧面的
lathe 车床,机床
layout 布局图,版面设计
lead alloy 铅合金
lead portion 料头
lead 螺距
lengthwise 纵向的
lenses 透镜
lettering 字体
limit switch 限位开关,行程开关
linear thermal expansion coefficient (LTEC) 线性热膨胀系数
load-bearing member 承载元
local forming 局部成形
locating pin 定位销
long fiber filler 长纤维填充剂
longitudinal 长的, 纵向的
low- and medium-volume production 中小量生产
low-melting 易熔的,低熔点的
low-pressure casting 低压铸造
lubricant 润滑剂
lubrication 润滑
lump 团,块
machinability 机械加工性,切削性
machinable 可加工的
machine cell 机加工单元
machine control unit 机床控制单元
machine leadscrew 机床丝杠
machine tool 机床

machine front 机器前端
machine-specific 指定的设备
machining 机械加工,切削加工
machining center 加工中心
magnesium alloy 镁合金
magnetic tape 磁带
magnitude 大小,数量,巨大
main core 主型芯
mainframe system 主机系统,大型机
manageability 可管理性
mandate 命令
manufacturing engineer 制造业工程师
manufacturing philosophy 制造原理
manufacturing system 制造系统
marinate 浸泡
mass production 大量生产
match-plate pattern 双面模板模
material data 材料参数
material database 材料数据库
materialize 物化
mathematical modeling 数学建模
mathematics 数学运算,数学应用
matrix 母体,基础
meat grinder 绞肉机
mechanism 机构,机理,机制
mechanization 机械化
melamine 三聚氰胺
melt temperature 熔化温度
melt-front 熔体前沿
melting point 熔点
mesh density 网格密度
mesh gauze 筛网过滤器
metal mold 金属模
metal processing 金属加工
metal projection 金属凸出部分
metal-cutting technique 金属切削技术
metallic engineering science 金属工程科学
metal-pumping system 金属浇注系统
metalworking machinery 金属加工机械
mica 云母
microcellular 微分网格
microchip encapsulation 微芯片封装
microcomputer 微型(电子)计算机
microprocessor 微处理器
microstructure 显微组织,微观结构
middle plate 中间板
midplane model 中型面模型
milling machine 铣床
mini-computer 小型机
miscellaneous function 辅助功能
mist coolant 喷雾冷却
module 模块
mold cavity 模具型腔
mold trial 试模
mold wall 模壁
moldability 可塑性
molding 造型,制模
molding compound 模塑料
molding cycle 成形周期
molecule 分子
molten state 熔融状态
monomer 单体
morphology 形态学
motor drive 电机驱动
mottle-colored part 斑点部分
mounting 安装,装配
moving core 活动型芯
moving core holder 活动型芯架
moving half of mold 瓣合机构
multicavity 多腔
multicavity center-gated 多腔中心浇口的
multiple process 复合过程
NC (numerical control) 数控
necking 缩颈
negative 负的,负数的
neutral axis 中性轴,中性层
nickel 镍
noncentral loading 偏心载荷
non-plastic 非塑性的
non-prototyping 非原型
nozzle 管口,喷嘴
nutshell flour 坚果壳粉
nylon 尼龙
ocean liner 远洋定期客轮
odd-shaped 奇形的
one-piece pattern 整体模

on-line assembly 在线装配
open riser 明冒口
operating system 操作系统
operation sequence 工序
optimization 最佳化
organic 有机的
orientation 位向,取向,定向
ornamental 观赏的
ornate 华丽的
orthotropic 正交各向异性的
overall effect 整体效果
overcut 过切
over-engineer 过加工
overflow well 溢流井
overhang 突出物,突出部分
overlap 重叠
overshearing 过剪切
oversimplified 简单化的
oxide layer 氧化层
oxidize 氧化
packing 填充,包装
packing pressure 保压压力
painting 喷漆
panel 面板
paper tape punch 纸带穿孔机
parabola 抛物线
parallel and coordinated fashion 并行协调方式
parison 型坯
part ejection 部分排除物
part family 零件族
part program 零/部件加工程序
part programmer 零件程序设计员
partial 部分的
parting 剖切,分开,分离
parting agent 脱模剂
parting line 分型线
parts list 零件目录表
patented fusion 独特的融合技术
pattern 模型
pellet 丸,粒
perforated 穿孔
perimeter 周长,周边
peripheral 边缘的,外表面的,周边的
permeability 渗透性
perpendicular to 垂直于
phenolic 酚醛塑料
phenotype 显型,表现型
photosensitive 感光的,光敏的
physical model 物理原型
physical setup 物理安装
pickling 酸洗
pigment 颜料
pillar die set 导柱模架
pillar guide 导柱
pipe extrusion 管材挤压
piston 活塞
pitch diameter 节圆直径
pitting 点蚀
plain linear motion 水平直线运动
planar triangle 平面三角形
plant designer 车间设计者
plastic deformation 塑性变形
plastic flow 塑性流动,塑性滑移
plastic forming 塑性成形
plasticate 塑炼,塑化(橡胶等)(指通过加热及挤压使其粒子软化)
plasticizer 可塑剂
plate 板,板材,钢板
platform 平台
plating 电镀
plotter 绘图机
plug point 插座
plumbing fitting 铅管制造装置
plumbing fixture 管道夹具
plunger 柱塞,滑阀
plunger bush 柱塞衬套
plunger-transfer 传递活塞
pocket 槽
Poisson's ratio 泊松比
polishing 抛光
polyamide 聚酰胺
polycarbonate 聚碳酸酯
polyethylene 聚乙烯
polymer 聚合体
polymeric resin 聚合树脂
polymerization 聚合

polypropylene 聚丙烯
polystyrene 聚苯乙烯
polyvinyl chloride 聚氯乙烯
porosity 多孔性,疏松度,孔隙度
positive 正的,正数的
post-processing 后处理
post-processor 后置处理程序
pouring 浇注
pouring basin 浇口杯
pouring cup 浇口杯,外浇口
powder 粉,粉末
powder-injection torch 喷粉焊炬
power tool 电动工具
precision-investment casting 精密熔模铸造
predetermined 预定的,预先确定的
pre-existing 预存
preheated perform 预热型坯
preliminary drawing 预制图
preload 预载,预加负载
premature 早期的
premium-quality 第一流的质量
preparatory function 准备功能
pre-processing 前处理
pre-processor 预处理操作
preset tool 预调刀具
press 冲压,压制
press platen 压板
press ram 压力机滑块
pressure gauge 压力计
pressure profile 压力分布图
printed circuit 印制电路
printer 打印机
priston-typed preplastifying machine 柱塞式预塑机
process chart 进程图
process controller setting 过程控制设置
process planner 过程计划者
process planning 制订工艺过程
process simulation 过程模拟
processing aid 加工助剂
processing workstation 处理站
processor 处理器
producing time 生产期
production run 生产过程
production-quality 生产
productivity rate 生产率
programmable automation 可编程自动控制
programmable logical controller 可编程逻辑控制器
progressive die 连续模
project 计划,工程
propagation 扩展
propeller 推进器
proprietary 专有的
prototype 原型
proven solution 证明方法
pulsating current 脉冲电流
pulse generator 脉冲发生器
pulsed 使产生脉冲
pump 泵
punch 冲头,冲孔
punched tape 打孔带
punching 冲孔,冲压
purge 清除,清洗
quadrant 象限,四分仪
quench 淬火
quenched 淬火
query 询问
quick-change die 快速凹模,快速换模
radii 半径
radiused 圆弧形的
ram-speed profile 滑块速度分布图
rapid prototyping 快速成形
rapid tooling 快速模具
rapid traverse 快速移动,快速行程
ratchet 棘轮
reaction injection 反应注射
recess 凹槽,凹处
reciprocate 互换
reciprocating-screw injecting machine 往复式螺杆注塑机
recur 重复,递归
register 寄存器
regrind material 再研磨材料
reinforcing 加强,强化
reject rate 废品率
relaxation 缓和,放松

re-machine　再加工
remainder　残余,剩余物
repetition　重复,循环
repetitive　反复的
reservoir　蓄水池,储水池
reshaping　整形
residual stress　残余应力
resin　树脂,涂树脂于
resolidification　再凝固作用
resultant casting　最终铸造
retraction　回复
retrieval　恢复,修补
reversal　反向,颠倒
rewind　倒回,重绕,回卷
rhenium　铼
rheological　流变的,流变学的
rheology　流变学
rigidity　刚性,刚度
rinsed　冲洗,漂洗
riser　冒口
riveting　铆接
robotics　机器人技术
robust　坚固的,健康的
roller feed　滚轮送料
rollover　塌角
rotary-tool　回转刀具
rotary-work　回转工件
rotating motion　旋转运动
rotating screw　旋转螺杆
rotating tool magazine　旋转式刀具库
rotational casting　离心浇铸
routing　选定路线
rubber injection molding　橡胶注射成形
runner plate　流道板
runner system　流道系统
runnerless molding　无流道模具
runners　分流道
saddle　鞍,滑板,座板
sand blasting　喷砂处理
sand casting　砂型铸造
sawing　锯,锯开
scheduling　编制目录
schematically　图表
scrap　废料,切屑
screen pack　过滤网
screw conveyor　螺旋输送机
sealing　密封
seam　缝,接缝
second machining　二次加工
secondary operation　二次加工
segregate　隔离
semifinished　半制成的
semipositive　半全压式
sequential　连续的,有顺序的
servo control　伺服控制
servo controlled feed　伺服控制进给
servo drive　伺服驱动
servo feed　伺服进给
servo-amplifier　伺服放大器
servo-drive unit　伺服驱动单元
severity　难度
shaded plot　阴影图
shakeout　抖掉,打型芯
shaving　修边,整修
shearing　剪切
shearing action　剪切作用
shear-thinning　剪切稀化
sheet plate forming　板料成形
sheet-metal parts　钣金零件
shipping　发货,装运
short fiber filler　短纤维填充剂
short shot　短射
shot　注射
shrinkage　收缩
shut　折叠
side product　副产品,废料
side-pull mechanism　侧拉机构
signal voltage　信号电压
silica sand　硅砂
silicon　硅
silicon chip　硅片
simple die　简单模
simplicity　简单,简易
simultaneously　同时的,同时发生的
sink mark　收缩痕
sintering　烧结

sizing 整形,矫正
skin layer 表层
skin material 表面材料
slice 薄片,切片
slide 滑块,滑板
sliding deformation 滑移变形
slippage 滑动
smoothness 光滑(度),平整(度)
sodium chloried 氯化钠
sodium nitrate 硝酸钠
softness 柔性,柔和
software 软件
software package 软件包
solid model 实体模型
solidification 凝固,固化
solver setting 解决方案设置
sophistication 混合
spacer 垫片
spacer block 垫块
spark machining 电火花加工
specially-shaped 异形的
spherical 球形的,圆的
spigot 塞子,栓
spin die 旋转模头
spindle 轴,主轴
spinning 旋压
split pattern 对分模,组合模
spring pad 弹簧垫
spring stripper 弹性卸料板
springback 回弹
spring-loaded 装弹簧的
sprue 浇道,主流道
squeeze 挤压,压印
straightness 直线度
stabilizer 稳定剂
stacked mold 重叠压塑模具
stagger 交错,错移
stamping 冲压,冲压件
standpoint 立场,观点
stationary pattern 固定台,固定板
stationary platen 固定压盘
stationary stripper 固定卸料板
steel shaft 钢杆
step-by-step command 步进命令
stereolithography 立体平版印刷
Stereolithography Apparatus 立体印刷成形
stiffness 刚性,刚度
stock 毛坯,坯料
stock guide 导料板,导料装置
stop pin 定位销,挡料销
storage 存储
straightening 校正
strap 带,皮带,用带捆扎
strength-to-weight ratio 强度重量比
stress-strain 应力应变
strip 条料,带料,脱模
stripper 脱模杆,卸料板
stroke length 行程长度
structural reaction injection molding (SRIM) 结构反应注射成形
submission 提交
substitute material 代用材料
subtract 减去,扣除
superimposed 叠加
support plate 支撑板
surface model 曲面模型
susceptibility 敏感性,磁化系数
susceptible 敏感的
sweep 扫描,清除,弯曲
sweeper 清洁装置
switched off 关闭
synchronize 同步,同时发生
synergy 协同,配合
synthetic 合成的
synthetic sand 合成沙
syrupy 像糖浆的,糖浆似的
syrupy liquid 黏流体
system display 系统展示
system software 系统软件
table 工作台
table salt 食盐
take into account 考虑,重视
take shape 成形
tangible 原型,有形的
tank 储水池
tap 攻螺纹

tape 线,带
tape reader 磁带播放机
taper cutting 锥度切割
teeth 齿
television set 电视接收机,电视机
temperature distribution 温度分布
temperature gradient 温度梯度
tensile 拉伸
tensile stress 张应力
tension 拉伸,拉力,张力
terminate 结束,终止
terminology 词汇,术语,专门名词
tetrafluoroethylene(Teflon) 聚四氟乙烯
tetrahedral 四面体的
thermal fatigue 热疲劳
thermal stability 热稳定性
thermocouple 热电偶
thermoforming 热成形
thermo-mechanical 热-机的
thermoplastics 热塑性塑料
thermoset plastics 热固性塑料
thin-walled part 薄壁制件
thread cutting 螺纹切削
three-dimensional forming 三维成形
three-plate mold 三板模
tilting 倾斜,翻转
tin alloy 锡合金
titanium aluminide 钛铝合金
toggle 套索钉,拴牢
tolerance 公差
tool changer 刀库
tool design 工具设计
tool handle 工具柄
torch 割炬,焊炬,喷管,切割器
torpedo 分料梭
torsion strength 转矩
transfer line 转换线
transfer molding 传递模塑法,连续自动送进成形
transfer sleeve 传递套筒
transformation 转变,变换,相变
transistor 晶体管
transition point 过渡点
transition software 转换软件
transparency 透明度
transverse contraction 横向收缩
transverse-flow 横流
traverse 横贯,横穿,横跨
trigonometry 三角学
trimming 切边,修边,修整
tryout 试验,检验
tubing 管道
tubular blank 管状坯
tubular electrode 管状电极
tubular-shaped 管状的
tungsten carbide 碳化钨
turbine blade 涡轮叶片
two-dimensional forming 二维成形
two-level molding 双层模塑
two-piece pattern 上下两件模
two-plate mold 双板模
ultraviolet 紫外线
ultraviolet inhibitor 紫外线抑制剂
unassigned 未定义,未分配,未赋值
unbending 不易弯曲的,松弛的
undercut 凹槽,倒拔模
underfill(flip-chip)encapsulation 封装
undertake 承担,采取
uniaxial 单轴,单向
unidirectional repeatability 单向重复性
uniform 统一的,一致的
unloading 卸料,卸载
unmanned operation 无人操作
unsurpassed 非常卓越的
urea 尿素
user 使用者
user program 使用程序
user-friendly environment 友好的用户界面
u-shaped structure U 形结构
utilization 利用
vacuum forming 真空成形
valence 化合价,原子价
valve 阀
vaporize 气化,蒸发
varnish 漆
vat 大桶
vee bending tool V 形弯曲模

velocity control　速率控制
velocity-to-pressure switch-over　速度－压力转换点
vent　出口，通风孔
vertical hydraulic press　垂直压力
vertice　顶点
viability　生存能力，发育能力
viscoelastic behavior　黏弹行为
viscoelasticity　黏弹性
viscosity　黏度
visual aids　直观教具
visualization　可视化
void　气孔，中空的
volatilization　挥发
voltmeter　电压计，电压表
volume mesh　体积网格
volumetric shrinkage　体积收缩量
volumetrically　容积的
vulcanization　硫化，硫化过程
warpage　翘曲
waspaloy　镍基合金
waterproof cover　防水罩
wearing　磨损
weld line　熔接痕
weld metal　焊接（焊缝）金属
welding　焊接
wind tunnel　风洞
wire frame　线框
wire frame　丝架
wire rest　运丝装置
wire spool　丝卷
wire-frame　线框
with respect to　关于，至于
wood flour　木粉
word address　字符地址，字符代码
workflow　工作流
workhead　工作夹具
working drawing　工作图
working portion of die　模具工作区
workplace　工作面，工作场所
workpart　工件
workpiece　工件
worn-out tools　破旧的工具
wrinkle　起皱
X-ray　X 射线
yield point　屈服点
yield strength　屈服强度，流动极限
Young's modulus　杨氏模量
zinc　锌
zinc-aluminum alloy　锌铝合金

References

[1] BLAZYNSKI T Z. Design of tools for deformation processes[M]. London: Elsevier Applied Science Publishers, 1986.

[2] KALPAKJIAN S,SCHMIND S R. Manufacturing Engineering and Technology-Machining[M]. 北京:机械工业出版社,2004.

[3] KALPAKJIAN S, SCHMIND S R. Manufacturing Engineering and Technology-Hot Processes[M]. 北京:机械工业出版社,2004.

[4] STOECKHERT K. Mold-Making Handbook[M]. Munich: Hanser Publisher, 1998.

[5] BOOTHROYD G. Product Design for Manufacture and Assembly[M]. Second Edition. New York: Marcel Dekker Incorporated, 2001.

[6] SUCHY I. Handbook of DieDesign[M]. Second Edition. New York :McGR AW-HILL, 2006.

[7] WALSH R A, CORMIER D R. McGraw-Hill Machining and Metalworking Handbook[M]. Third Edition. New York:R. R. Donnelley & Sons Company, 2005.

[8] CHANDA M, ROY S. Plastics Technology Handbook [M]. Fourth Edition. New York: CRC Press,2006.

[9] CHEREMISINOFF NICHOLAS P. Product Design and Testing of Polymeric Materials[M]. New York: M. Dekker, c, 1990.

[10] BECK, RONALD D. Plastic product design[M]. New York: Van Nostrand Reinhold Co. , 1980.

[11] RAHM, LOUIS F. Plastic molding: an introduction to the materials, equipment and methods used in the fabrication of plastic products[M]. New York, London: McGraw-Hill book company, inc. ,1933.

[12] WALLER J A. Press Tools and Presswork[M]. Great Britain: Portcullis Press Ltd. , 1978.

[13] DRIVER, WALTER E. Plastics Chemistry and Technology[M]. New York: Van Nostrand Reinhold, 1979.

[14] BAIJAL, MAHENDRA D. Plastics Polymer Science and Technology[M]. New York: Wiley, 1982.

[15] THRONE, JAMES L. Plastics Process Engineering[M]. New York: M. Dekker, 1979.

[16] LEVY, SIDNEY, DUBOIS, HARRY J. Plastics Product Design Engineering Handbook[M]. New York: Chapman and Hall, 1984.

[17] MILLER, EDWARD. Plastics Products Design Handbook, Processes and Design for Processes[M]. New York: Marcel Dekker, Inc. ,1983.

[18] WHELAN A. Injection Moulding Machines[M]. London: Elsevier Applied Science, 1984.

[19] Verein Deutscher Ingenieure. Injection Moulding Technology. Dusseldorf: [s. l.], 1981.

[20] American Society for Metals, Casting Design Handbook[M]. Metals Park, Ohio. New York: Reinhold Pub. Corp, 1962.

[21] BUTLER J. Compression and transfer moulding of plastics[M]. London: Iliffe & Sons, 1959.

[22] American Foundrymen's Society. Molding Methods and Materials. Des Plaines, 1962.

[23] 颜寿葵,周大隽. 英汉锻压科技词汇[M]. 北京:地震出版社,1995.

[24] 王晓江. 模具设计与制造专业英语[M]. 北京:机械工业出版社,2004.

[25] 张晓黎,李海梅. 塑料加工和模具专业英语[M]. 北京:化学工业出版社,2005.

[26] 胡占齐,等. 数控技术[M]. 武汉:武汉理工大学出版社,2004.

[27] 王同亿. 英汉科技词汇大全[M]. 北京:科学普及出版社,1984.

[28] 《新英汉词典》编写组. 新英汉词典[M]. 上海:上海译文出版社,1982.

[29] 夏琴香. 冲压成形工艺及模具设计[M]. 广州:华南理工大学出版社,2004.

[30] 曹同玉,冯连芳. 高分子材料工程专业英语[M]. 北京:化学工业出版社,1999.
[31] 黄义俊. 数控技术专业英语[M]. 北京:人民邮电出版社,2009.
[32] 黄星. 数控技术专业英语[M]. 北京:机械工业出版社,2009.
[33] 屈利刚. 先进制造技术专业英语阅读[M]. 北京:化学工业出版社,2006.
[34] (美)拉奥. 制造技术(第 2 卷金属切削与机床英文版原书第 2 版)[M]. 北京:机械工业出版社,2010.
[35] 吴凤仙. 数控英语(英汉对照)[M]. 武汉:武汉大学出版社,2008.
[36] 刘振康. 机械制造英语读本[M]. 北京:机械工业出版社,1988.